D1085211

THE

California

ENERGY CRISIS

Lessons for a
Deregulating Industry

THE

California

ENERGY CRISIS

Lessons for a
Deregulating Industry

Will McNamara

PennWell®

Copyright © 2002 by
PennWell Corporation
1421 S. Sheridan Road
Tulsa, Oklahoma 74112
800-752-9764
sales@pennwell.com
www.pennwell-store.com
www.pennwell.com

cover design by Robin Brumley
cover illustration and book design by Amy Spehar

Library of Congress Cataloging-in-Publication Data Pending

McNamara, Will,
     The California Energy Crisis: Lessons for a Deregulating Industry /
Will McNamara.
          p. cm.
     Includes Index
     ISBN 0-87814-844-2

Printed in the United States of America

1   2   3   4   5   06   05   04   03   02

Dedication

This book is dedicated to citizens of California, who became participants in the first electric power industry deregulation experiment in the United States. Their experience, while tremendously challenging, has provided invaluable lessons for other states that proceed down the path of electric power industry restructuring.

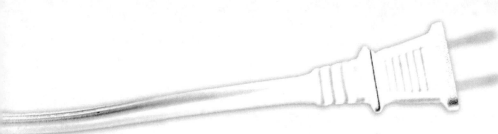

Contents

Acknowledgements

This book was written in the stimulating work environment of SCIENTECH, Inc., where a group of exceptional professionals devote their careers to the study of trends within the evolving energy industry and to the application of practical solutions to challenges within this market.

My gratitude is extended to all of my colleagues within this organization for the intellectual exchange and support they have provided with regard to my study of U.S. energy markets. I am particularly grateful to Robert C. Bellemare, P.E., vice president of SCIENTECH, who served as my mentor and sounding board as I formed impressions and evaluations of California's experience with electric power deregulation.

Introduction
Part I
California's Energy Dreams

CALIFORNIA, SEPTEMBER 20, 2001: DIRECT ACCESS IS TERMINATED

It seemed strangely anticlimactic, but monumental nonetheless. The five members of the California Public Utilities Commission (CPUC) issued a 3 to 2 vote that signaled the final death sentence for electric power competition (otherwise known as direct access) in the state of California. The vote terminated what had been a massive undertaking by a variety of state and federal stakeholders to give Californians choice over how and where their power originated. Yet, the vote also attempted to close the book on what had been generally thought to be a colossal mess. Amid bankrupt and financially paralyzed utilities, rolling power outages, and a state economy that had been severely compromised, the CPUC acknowledged that electric power competition in California had failed. The program that had begun with an admirable intention—lowering sky-high electricity rates for California residents and businesses—had arguably caused more trouble than benefits for its many participants.

Yet, one nagging question remained atop a long list of other conundrums: What on earth had gone wrong? How could a program that had been planned and implemented with great confidence by the state's legislators, regulators, utilities, and consumer advocates come to such a disastrous end? While those questions have many diverse answers, one thing is crystal clear. The electric power choice experiment in California that has had such a checkered history is now officially dead, and the state faces the arduous task of putting the "deregulation genie" back into its bottle.

Of course, direct access was never what could be considered a success in California. Since its official launch on March 31, 1998, most Californians shunned the opportunity to switch to a new electric power provider and opted to stay with one of the three incumbent investor-owned utilities (IOU) in the state—Pacific Gas & Electric Co. (PG&E), Southern California Edison (SCE), and San Diego Gas & Electric Co. (SDG&E)—which collectively provide power to the vast majority of the state's residents. At its peak in 1998, direct access in California enticed about 2.5% of residential customers and 13% of industrial customers to contract with a new power supplier. Over the last year, this number has dropped dramatically as contracts have expired. It is arguable that direct access was somewhat successful for large industrial customers in California, who perhaps gained some level of measurable savings by switching to a new provider. However, as a whole, the CPUC estimates that only 5% of California's peak electricity demand of 46,000 MW was in direct access contracts when the commission voted to terminate electric power choice in the state. Thus, if the CPUC's vote truly represented a death sentence for California deregulation, then it must be conceded that the state's energy market had been on life support for some time.

Nevertheless, in retrospect, it was no surprise that California was in fact the first U.S. state to jump on the electric power restructuring bandwagon. After all, California's electricity rates traditionally had been 30–50% above the national average, exceeded only by rates in New England (not surprisingly, the other region of the country where deregulation has proceeded forward most expeditiously) and in Hawaii. California residents historically had paid electricity rates double and triple those of other states. And California's large corporations, specifically industrial and agricultural companies, had been especially impacted by the state's high electricity costs.

Thus, direct access was created to eradicate the monopoly system among incumbent electric power utilities and hopefully bring lower rates to electricity customers through the onset of choice of power suppliers. As the examination of California's "competitive" electric power market in this text will show, deregulation in the state had such disastrous impacts that the state government felt forced to step in and assume control over key parts of the businesses, traditionally operated by the three IOUs. This assumption of control of the energy market by the state government has potentially put California and all of its citizens at financial risk for years to come.

Specifically, since the California Department of Water Resources assumed the role of operating as California's primary power purchasing agent, as part of Gov. Gray Davis' efforts to gain more control over the state's erratic energy market, the state government has already amassed debts over $9 billion (as of this writing). It is the state's intent to recoup this debt through rates affixed on energy customers served by the three incumbent utilities. Yet, had direct access been kept alive, it would have kept the door open for large energy customers to sign deals with other power suppliers and thus avoid paying any share of the state's energy tab, including purchases already made. This theoretically could have left the state with a surplus of costly electric power and few customers to pay for it.

In addition, the state feared that if small consumers were left to pay a disproportionately high amount of the state's debt due to larger customers leaving the system, this would have resulted in a mass exodus from California, which would also have had a negative impact on the state's stability. This dynamic is known as the "fixed-cost death spiral," which the state is now trying to avoid. In the words of Loretta Lynch, the CPUC's president, "We need to ensure a customer base...so that as the power is purchased by the customer, the state is repaid." Consequently, despite the fact that electric competition was not a success, the state actually opted to discontinue direct access as a way to protect the economy of the state government as a whole. Further, now that California's direct access has been officially discontinued, the policy will most likely remain in place for at least 10 years, or through the tenure of the long-term contracts that the state has signed with various suppliers.

Nevertheless, the state of California has assumed an enormous and risky role as it not only has become committed to long-term (and costly) power buying contracts, but also seeks to purchase transmission assets from incumbent utilities and expropriate generation assets. Thus, even though direct access is dead, the state of California is still encountering significant residual effects from the failed experiment of electric power competition. In addition to the fact that the California government is arguably not prepared to manage the intricacies of the state's complex and huge electric power market, there are substantial financial risks associated with the government's new role.

For instance, although the actual requirements of the long-term power deals have remained proprietary, it is a fairly safe bet that the 10-year contracts the state of California has entered into include energy prices of about seven cents/kWh. This average compares to wholesale prices of three to four cents/kWh that were common in the mid- to late-1990s. If prices return to the three to four cents/kWh range or anything below the 7 cents/kWh range, the state of California will remain locked in long-term contracts based on high prices. These high prices will create a disproportionately high amount of debt for the state government and possibly excess electric power that the state will not be able to sell at a profit. Granted, the CPUC now has unilateral authority to increase rates to compensate the state for power purchases, but conflict would certainly result if California's rates continue to exceed market rates.

So, what does all of this mean? The CPUC has likely based its decision to end direct access on a belief that "desperate circumstances call for desperate measures." However, the desperate circumstances in which California still finds itself resulted from bad decisions made in the planning of the state's deregulated electric power market, including decisions made by the CPUC. Moreover, the decision from the CPUC furthers the transfer of financial vulnerability from the incumbent utilities to the state of California, which creates a very precarious situation for the nation's largest energy market. In addition, electric power competition in the nation's largest state, which represents 20% of the country's economy, will now be blocked for at least the next decade. This potentially could have a very damaging effect on the level of new technologies and industries that might be brought into (or choose to remain in) the state of California. The decision could also have tremendous impact on other states across the nation that have yet to implement any variation of electric power competition. The total consequences of this measure remain to be seen, but it is clear that the CPUC's decision to end direct access is a huge milestone in the ongoing California energy industry debacle.

Clearly, California is often viewed as a trendsetter for the rest of the United States and, increasingly, the world. Usually, things that happen in the Golden State will happen throughout the rest of the world five or six years later. As such, all eyes that follow the energy industry as it deregulates (albeit sporadically) across the United States continue to watch California as an indicator for what might happen elsewhere. The issue of whether or not California represents a template for how deregulation will unfold across the United States is a loaded question, with different answers depending on who is responding.

Yet, despite many different interpretations and explanations, there are some empirical facts that can be gleaned about the California experiment with electric power deregulation and the various factors that collided to cause the experiment to fail. Based on facts, this text will attempt to provide one comprehensive theory as to the root causes of the California energy crisis and the various pros and cons associated with the efforts to resolve the crisis. The theoretical questions that will be addressed and answered in this book are as follows:

- What were the problems that caused California's competitive energy market to fail?
- What steps could have been taken to create a more successful outcome?
- Who is to blame for the California energy crisis?
- What does the future hold for California's energy market?
- How has the failure of direct access in California impacted other states and the country as a whole?
- Does the California electric power deregulation experience represent a template for how electric power competition will unfold in other states?

Introduction Part II
Main Players in the California Energy Crisis

The California energy crisis involved a cast of diverse participants that all had unique agendas and objectives. Throughout this text, I highlight the various participants and the role that they played both in the creation of the energy crisis of the state and the attempts made toward its resolution.

STATE OF CALIFORNIA PARTICIPANTS

There are three investor-owned utilities (IOU) in California, which together deliver 70% of the state's power. The three IOUs were most directly impacted by the massive electric power restructuring program that took place in California, which began March 31, 1998.

California Investor-Owned Utilities
PG&E Corporation / Pacific Gas & Electric Co.
PG&E Corporation

With revenues of almost $21 billion, PG&E Corp. markets energy services and products throughout North America through its National Energy Group. PG&E Corp.'s businesses also include PG&E Co., the Northern and Central California utility that delivers natural gas and electricity service to one in every 20 Americans. Key information about PG&E Corp. that is important to note includes the following:

- Operations in 21 states
- One of the largest U.S. transporters of Canadian natural gas
- 30 power plants in operation, two plants under construction
- Generation portfolio of 7,000 MW
- More than 10,000 MW in new power plant development and construction
- Owners of one of the largest gas and electric utilities in the country
- NYSE stock symbol: PCG
- 151,000 common shareholders

Pacific Gas & Electric Co.

PG&E Co., incorporated in California in 1905, is one of the largest combination natural gas and electric utilities in the United States. Based in San Francisco, the utility serves 13 million people throughout a 70,000-square-mile service area in Northern and Central California

- Service area stretches from Eureka in the north to Bakersfield in the south, and from the Pacific Ocean in the west to the Sierra Nevada in the east.
- 131,000 circuit miles of electric power lines
- 43,000 miles of natural gas pipelines

- 4.5 million electricity customer accounts
- 3.7 million gas customer accounts

PG&E Co. declared bankruptcy in April 2001 as a result of huge debts it carried related to the escalating price of wholesale electric power and its inability to pass on those costs to customers due to a statewide retail rate freeze. As of this writing (December 2001), the utility continues to develop a restructuring plan as it navigates through bankruptcy proceedings.

Edison International/Southern California Edison
Edison International

Edison International is an international electric power generator, distributor, and structured finance provider. Edison's family of closely related companies develops, acquires, finances, owns, operates, and maintains reliable and efficient electric power systems worldwide.

- Power generation portfolio of about 28,000 MW
- Operations in nine countries and four regions of the United States
- Combined assets of more than $36.9 billion
- Headquarters in Rosemead, California

Southern California Edison (SCE)

SCE is one of the largest electric utilities in the United States, and the largest subsidiary of Edison International. On an average day, SCE provides electric power for 800 communities and cities, 5,000 large businesses, and 280,000 small businesses in Central and Southern California. Delivering that electric power takes 16 utility interconnections, 4,900 transmission and distribution circuits, 365 transmission and distribution crews, the days and nights of 12,642 employees, and over a century of experience. Excluding the city of Los Angeles, which is served by the municipal utility known as the

Los Angeles Department of Water and Power (LADWP), SCE's 50,000 square-mile service territory has a population of 11 million persons.

Sempra Energy/San Diego Gas & Electric
Sempra Energy

Sempra Energy is a global corporation uniquely positioned to take advantage of the dramatic changes reshaping energy markets worldwide. The 1998 merger of Pacific Enterprises and Enova Corporation enabled Sempra Energy to begin its new mission with a powerful presence in the energy services industry worldwide and to provide the most innovative and efficient energy solutions possible. Based in San Diego, Sempra Energy is a Fortune 500 company with eight subsidiaries.

San Diego Gas & Electric

Founded in 1881, SDG&E delivers electricity to 18 cities in San Diego County and seven cities in the southernmost section of Orange County. SDG&E is a regulated distribution utility providing electric service to three million customers in San Diego and southern Orange counties and natural gas service to San Diego County.

The California Independent System Operator (ISO)

The California power grid is a network of long-distance, high-voltage transmission lines and substations that carry bulk electricity to local utilities for distribution to their customers. The mission of the California ISO is to safeguard the reliable delivery of electricity and ensure equal access to the power grid. Toward that end, the California ISO controls 75% of California's power grid, transmission systems formerly operated by the three investor-owned utilities in the state.

The California ISO-controlled portion of the grid covers 124,000 square miles or three-quarters of the state. Power plants meeting up to 45,000 MW of peak demand are connected to the California ISO grid, making the control area the second largest in the United States (Pennsylvania-New Jersey-Maryland Interconnection is the largest) and the fifth largest in the world. Part of California ISO's commitment to reliability requires a transmission-planning role. California ISO will plan grid enhancements to meet its high standards for reliability, at a minimum cost to consumers.

Transmission owners (predominantly IOUs in the California market) file annual transmission expansion plans to accommodate the state's growing electricity needs. The California ISO reviews and either approves or makes recommendations regarding the proposed additions. Recommendations that are not accepted proceed to dispute resolution. As part of a Coordinated Planning Process, the Cal-ISO will work with Regional Transmission Groups (RTG) and the Western Systems Coordinating Council (WSCC) to ensure expansion projects do not negatively impact the regional grid and transmission owners in other states.

California PX

No longer operational. In operation from April 1998 until early 2001, the California PX was an independent entity that set the price of electricity in the state of California depending on a continually fluctuating balance of supply and demand. The price of electricity within the California PX system was essentially the weighted average of the day-ahead and hour-ahead prices, as calculated by various bids entered by wholesale generators. The California PX ceased to exist when the Department of Water Resources, a state agency, stepped into the role as California's primary electric power purchaser under the authority of Gov. Gray Davis in January 2001.

California Energy Commission (CEC)

The CEC is the state's primary energy policy and planning agency, charged with ensuring a reliable and affordable energy supply. The Commission has five major responsibilities:

- Forecasting future energy needs and keeping historical energy data;
- Siting and licensing power plants;
- Promoting energy efficiency through appliance and building standards;
- Developing energy technologies and supporting renewable energy; and
- Planning for and directing state response to energy emergencies.

With former Gov. Pete Wilson's signing of the Electric Industry Deregulation Law (AB 1890), the CEC's role includes overseeing funding programs that support public interest energy research; advancing energy science and technology through research, development, and demonstration; and providing market support to existing, new, and emerging renewable technologies.

The governor of California appoints, with Senate confirmation, five commissioners to staggered five-year terms. The commissioners must come from and represent specific areas of expertise: law, environment, economics, science/engineering, and the public at large. Current commissioners as of December 2001 are the following:

- Chair William J. Keese
- Commissioner Michael C. Moore, Ph.D.
- Commissioner Robert A. Laurie
- Commissioner Robert Pernell
- Commissioner Arthur H. Rosenfeld, Ph.D

The California Public Utilities Commission (CPUC)

The CPUC regulates privately owned telecommunications, electric power, natural gas, water, railroad, rail transit, and passenger transportation companies. According to its Web site, the CPUC is responsible "for assuring

California utility customers have safe, reliable utility service at reasonable rates, protecting utility customers from fraud, and promoting the health of California's economy." In pursuing these goals, the Commission establishes service standards and safety rules, and authorizes utility rate changes. It monitors the safety of utility and transportation operations, and oversees markets to inhibit anti-competitive activity. In its efforts to protect consumers, it prosecutes unlawful utility marketing and billing activities, governs business relationships between utilities and their affiliates, and resolves customer complaints against utilities. It implements energy efficiency programs, low-income rates, and telecommunications services for disabled customers. It oversees the merger and restructure of utility corporations, and enforces the California Environmental Quality Act for utility construction. The CPUC works with other state and federal agencies in promoting water quality, environmental protection, and safety. It also intervenes in federal proceedings on issues that affect California utility rates or services.

The Governor appoints the five commissioners, who must be confirmed by the Senate, for six-year staggered terms. The Governor appoints one of the five to serve as commission president. The CPUC's headquarters are in San Francisco.

The CPUC's current commissioners, as of December 2001, are the following:

- President Loretta Lynch (appointed Feb. 5, 2000, by Gov. Gray Davis)
- Commissioner Carl Wood (appointed June 21, 1999, by Gov. Gray Davis)
- Commissioner Geoffrey Brown (appointed Jan. 18, 2001, by Gov. Gray Davis)
- Commissioner Henry M. Duque (appointed April 3, 1995, by former Gov. Pete Wilson)
- Commissioner Richard A. Bilas (appointed Jan. 1, 1997, by former Gov. Pete Wilson)

Past CPUC commissioners who played a role in the creation of California's direct access program were the following:

- P. Gregory Conlon (appointed Feb. 11, 1993, by former Gov. Pete Wilson; served as president of the CPUC from May 1996 to December 1997)
- Jessie J. Knight, Jr. (appointed Aug. 23, 1993, by former Gov. Pete Wilson)
- Norman Shumway (appointed Feb. 12, 1991, by former Gov. Pete Wilson; resigned Feb. 23, 1995)
- Daniel Fessler (appointed Feb. 12, 1991; served as president from 1992 to April 1996)
- Patricia Eckert (appointed March 13, 1989; served as president in 1991)
- Joel Hyatt (appointed June 10, 1999; resigned)

California Gov. Gray Davis

Joseph Graham Davis, Jr. (nicknamed Gray by his mother) was overwhelmingly elected the 37th Governor of California on November 3, 1998, winning 58% of the statewide vote. In the June primary election, Davis shocked political observers by not only handily defeating two better-funded Democratic opponents, but by also finishing ahead of the unopposed Republican nominee. It was the continuation of an old tradition; in his successful campaign for Lieutenant Governor in 1994, he received more votes than any other Democratic candidate in America.

The California Assembly

The following information was taken from the Official California Legislative Information Web site (www.http://www.leginfo.ca.gov/).

Overview of the Legislative Process in California

The California State Legislature is made up of two houses: the Senate and the Assembly. There are 40 Senators and 80 Assembly Members representing the people of the State of California. The Legislature has a legislative calendar containing important dates of activities during its two-year session. The legislative process by which a bill becomes a law in California includes the following steps.

1. **Idea** – All legislation begins as an idea or concept. Ideas and concepts can come from a variety of sources. The process begins when a Senator or Assembly Member decides to author a bill.

2. **The Author** – A Legislator sends the idea for the bill to the Legislative Counsel, where it is drafted into the actual bill. The draft of the bill is returned to the Legislator for introduction. If the author is a Senator, the bill is introduced in the Senate. If the author is an Assembly Member, the bill is introduced in the Assembly.

3. **First Reading/Introduction** – A bill is introduced or read the first time when the bill number, the name of the author, and the descriptive title of the bill is read on the floor of the house. The bill is then sent to the Office of State Printing. No bill may be acted upon until 30 days have passed from the date of its introduction.

4. **Committee Hearings** – The bill then goes to the Rules Committee of the house of origin where it is assigned to the appropriate policy committee for its first hearing. Bills are assigned to policy committees according to subject area of the bill. For example, a Senate bill dealing with healthcare facilities would first be assigned to the Senate Health and Human Services Committee for policy review. Bills that require the expenditure of funds must also be heard in the fiscal committees: Senate Appropriations or Assembly Appropriations. Each house has a number of policy committees and a fiscal committee. Each committee is made up of a specified number of Senators or Assembly Members.

   During the committee hearing, the author presents the bill to the committee, and testimony can be heard in support of or opposition to the bill. The committee then votes by passing the bill, passing the bill as amended, or defeating the bill. Bills can be amended several times. Letters of support or opposition are important and should be mailed to the author and committee members before the bill is scheduled to be heard in committee. It takes a majority vote of the full committee membership for a bill to be passed by the committee.

Each house maintains a schedule of legislative committee hearings. Prior to a bill's hearing, a bill analysis is prepared that explains current law, what the bill is intended to do, and some background information. Typically, the analysis also lists organizations that support or oppose the bill.

5. **Second and Third Reading** – Bills passed by committees are read a second time on the floor in the house of origin and then assigned to third reading. Bill analyses are also prepared prior to third reading. When a bill is read the third time, it is explained by the author, discussed by the members, and voted on by a roll call vote. Bills that require an appropriation or that take effect immediately generally require 27 votes in the Senate and 54 votes in the Assembly to be passed. Other bills generally require 21 votes in the Senate and 41 votes in the Assembly. If a bill is defeated, the member may seek reconsideration and another vote.

6. **Repeat Process in other House** – Once the bill has been approved by the house of origin, it proceeds to the other house, where the procedure is repeated.

7. **Resolution of Differences** – If a bill is amended in the second house, it must go back to the house of origin for concurrence, which is agreement on the amendments. If agreement cannot be reached, the bill is referred to a two-house conference committee to resolve differences. Three members of the committee are from the Senate and three are from the Assembly. If a compromise is reached, the bill is returned to both houses for a vote.

8. **Governor** – If both houses approve a bill, it then goes to the Governor. The Governor has three choices. The Governor can sign the bill into law, allow it to become law without his or her signature, or veto it. A governor's veto can be overridden by a two-thirds vote

in both houses. Most bills go into effect on the first day of January of the next year. Urgency measures take effect immediately after they are signed or allowed to become law without signature.

California Law

Bills that are passed by the Legislature and approved by the Governor are assigned a chapter number by the Secretary of State. These Chaptered Bills (also referred to as Statutes of the year they were enacted) then become part of the California Codes. The California Codes are a comprehensive collection of laws grouped by subject matter.

The California Constitution sets forth the fundamental laws by which the State of California is governed. All amendments to the Constitution come about as a result of constitutional amendments presented to the people for their approval.

Department of Water Resources

The California state agency that assumed the role of primary power purchaser for California's three IOUs in January 2001 when the credit standing of the IOUs became questionable.

FEDERAL/INTERSTATE PARTICIPANTS

The Federal Energy Regulatory Commission (FERC)

FERC is an independent regulatory agency within the Department of Energy that performs the following tasks:

- Regulates the transmission and sale of natural gas for resale in interstate commerce;
- Regulates the transmission of oil by pipeline in interstate commerce;
- Regulates the transmission and wholesale sales of electricity in interstate commerce;
- Licenses and inspects private, municipal, and state hydroelectric projects;
- Oversees environmental matters related to natural gas, oil, electricity, and hydroelectric projects;
- Administers accounting and financial reporting regulations and conduct of jurisdictional companies; and
- Approves site choices as well as abandonment of interstate pipeline facilities.

The Commission recovers all of its costs from regulated industries through fees and annual charges.

FERC was created through the Department of Energy Organization Act on October 1, 1977. At that time, the Commission's predecessor, the Federal Power Commission (FPC), was abolished, and the new agency (FERC) inherited most of the FPC's responsibilities.

The Commission's legal authority comes from the Federal Power Act of 1935, the Natural Gas Act (NGA) of 1938, the Natural Gas Policy Act (NGPA) of 1978, the Public Utility Regulatory Policies Act (PURPA) of 1978, and the Energy Policy Act of 1992.

The Commission is composed of five members who are appointed by the President of the United States, with the advice and consent of the Senate. Commissioners serve five-year terms, and have an equal vote on regulatory matters. No more than three members may belong to the same political party. One member is designated by the President to serve as Chair, and FERC's administrative head. The Commission's current membership consists of Pat Wood III (Chairman), William L. Massey, Linda Key Breathitt, and Nora Mead Brownell.

In October 2001, President Bush nominated electricity expert Joseph Kelliher to fill the fifth spot on FERC. Kelliher, a Republican, is well known in Washington energy circles. He was a member of the Bush transition team during the winter of 2000-2001, an aide to Texas Rep. Joe Barton on electricity deregulation, and most recently a top advisor to Energy Secretary Spencer Abraham. The U.S. Senate must confirm the nomination of Kelliher before he can be sworn in as a commissioner.

Western Systems Coordinating Council (WSCC)

Western Systems Coordinating Council (WSCC) was formed with the signing of the WSCC Agreement on August 14, 1967 by 40 electric power systems. Those "charter members" represented the electric power systems engaged in bulk power generation and/or transmission serving all or part of the 14 Western States and British Columbia, Canada. Membership in WSCC is voluntary and open to major transmission utilities, transmission-dependent utilities, and independent power producers/marketers. In addition, affiliate membership is available for power brokers, environmental organizations, state and federal regulatory agencies, and any organization having an interest in the reliability of interconnected system operation or coordinated planning.

The WSCC region encompasses a vast area of nearly 1.8 million square miles. It is the largest and most diverse of the 10 regional councils of the North American Electric Reliability Council (NERC). WSCC's service territory extends from Canada to Mexico. It includes the provinces of Alberta and British Columbia, the northern portion of Baja California, Mexico, and all or portions of the 14 western states in between. Transmission lines span long distances, connecting the verdant Pacific Northwest with its abundant hydroelectric resources to the arid Southwest with its large coal-fired and nuclear resources.

Wholesale Power Generators

When direct access began in California, wholesale power generators were able to begin the marketing and trading of electric power in the state's market. There are eight companies that later became known as having the largest trading volumes in the state of California. In addition, as of this writing, these companies face continued investigation by FERC for potential market manipulation and may have to issue refunds to the state of California:

Arlington, VA-based AES Corp. (NYSE: AES)
San Jose, CA-based Calpine (NYSE: CPN)
Charlotte, NC-based Duke Energy (NYSE: DUK)
Houston, TX-based Dynegy Inc. (NYSE: DYN)
Houston, TX-based Enron Corp. (NYSE: ENE)
Houston, TX-based Reliant Energy (NYSE: REI)
Atlanta, GA-based Mirant (NYSE: MIR)
Tulsa, OK-based Williams Companies (NYSE: WMB)

Chapter 1
Historical Perspective

Before any substantive discussion of the causes of what has been collectively referred to as the California energy crisis can take place, it is important to understand the foundation for electric power competition that was established even before the crisis began. In other words, the crisis in California's energy market did not emerge in a vacuum, but rather was a manifestation of an array of decisions made by various parties as the state formulated its restructuring plan.

It would be very difficult to comprehend the complex dynamics that led to California's energy crisis without first establishing the larger market conditions under which the state's competitive electric power market came to be. Moreover, causal factors that developed before the launch of electric power competition in California in April 1998 are crucial to understanding the events of today and tomorrow.

THE EARLY DRIVERS OF DEREGULATION

Deregulation of the U.S. energy industry has roots that extend 20 years back. In response to the 1973 Oil Embargo of the Arab-Israeli war, the National Energy Act created a new class of non-utility electric power generators and also deregulated the prices paid to gas producers. Previously, prices had been governed by complex regulations that constrained supplies. The Act put into motion further measures that gradually deregulated the natural gas industry. Electric power deregulation started later, but is occurring at a faster pace (at least in some areas of the country), and of course California would eventually become the first U.S state to dive headfirst into electric power competition.

However, prior to 1976, California gave little thought to electric power competition or deregulation, and during this time the state's investor-owned utilities maintained control over the production and distribution of electricity to the regions that they served. Under this long-standing monopoly structure, the roles were very clear. The state's three incumbent utilities owned the power plants, transmission systems, and distribution lines needed to produce and deliver electric power to California homes and businesses. The California Legislature gave leverage to the utilities to earn what was considered a fair profit margin in their business, and empowered the CPUC to plan for long-term electric power needs and oversee the utilities' rate structures and operations.

In 1976, the state of California, ever the pioneer, made its first initial break from the long-standing monopoly system when it enacted legislation to encourage private energy producers to develop sources of non-fossil generated electric energy. These new forms of power supply included wind energy, solar, and other forms of renewable power, which rode the wave of interest in these environmentally friendly power sources, that was common in California during this time. The new legislation in California and extension of California's electric power market away from the heavy influence of the incumbent utilities are significant to subsequent events. It encouraged other power companies to develop new sources of energy supply away from the utilities, and required the utilities to provide access for these independent companies across the state's transmission lines.

Public Utility Regulatory Policies Act (PURPA)

A major inclusion of the National Energy Act was PURPA of 1978, enacted by President Jimmy Carter. The primary goal of this law was to lower electricity costs by forcing utilities to purchase electric power from a range of independent power producers. The law was designed to encourage use of domestic electricity sources—natural gas, coal, and renewable energy sources. PURPA also encouraged energy conservation and efficiency, and more "equitable" electric utility rates. PURPA required utilities to buy power from unregulated generators at "avoided costs" prices to encourage the growth of electricity supplies and alternative energy. These new generators, commonly referred to as qualifying facilities (QF), used co-generation or other environmentally preferred technologies, such as solar, wind, or biomass. The law also allowed these independent producers to sell directly to large industrial customers and utilities, which signaled another departure from the monopoly system. Following the lead of this federal policy, the CPUC began to develop QF standards for California. By 1984, QF projects totaled more than 10,000 MW in California, which was either online or had signed delivery contracts to the utilities.

As a result of PURPA, the CPUC began in 1978 to develop standards under which the IOUs would be required to purchase power from the QFs. Concurrently, the California Energy Commission set a goal to reduce the proportion of the state's electricity generated by oil and natural gas to no more than 33% by 1990.

Since PURPA was passed in 1978, the national market for new generation capacity has been essentially deregulated and has become increasingly competitive. PURPA was a very important law because it introduced competition into the power generation marketplace and demonstrated that competition could work and should be allowed to work. Nevertheless, the impact that PURPA had in California was to bring a sense of uncertainty about the impact that competition in the generation side of the energy would have on incumbent utilities. As a result, the incumbent utilities were even more reticent to invest money in additional generating capacity, and the period of 1990-1996 in California was marked by a considerable slowdown of additional plant development.

Consequently, very little new generation was proposed in the state, especially during the period 1985 to 1990. Conservation programs and the availability of low-cost surplus power from neighboring regions such as the Northwest and Southwest installed a false sense of adequate generation supply in California. Specifically, the Southwest had overbuilt its own electricity supply system with coal-fired generating plants to meet a projected energy demand that did not materialize, and thus had surplus power to sell to California. Due to this excess in supply, California was able to buy power from the Southwest at an average price of one cent/kWh. This would be another factor that eventually would later emerge as a leading cause of the California energy crisis.

The National Energy Policy Act

The National Energy Policy Act of 1992 was signed into law by President George Bush. The Act marked the end of a highly structured and regulated period for investor-owned utilities and the beginning of a new age characterized by intense competition. This competition, which is still continuing to unfold 10 years later, is driven by rival economic forces and encouraged by the open-market regulatory practices of the FERC.

Perhaps most importantly, the National Energy Policy Act gave transmission grid access to independent power producers, enabling a new market for merchant generators to emerge. In other words, the monopoly system that existed for decades was further dismantled by the National Energy Policy Act, which allowed private, for-profit power generating companies to sell electricity on the open market and across the transmission systems owned by incumbent utilities. The Act also deregulated the natural gas market.

It is also important to note that the National Energy Policy Act accomplished the following:

- Established a new class of independent power producers known as exempt wholesale generators;
- Allowed the FERC to order utilities to transmit or "wheel" power produced by independent producers;
- Prohibited the FERC from ordering electric utilities to engage in retail wheeling; and
- Permitted electric utility holding companies to acquire and maintain financial interests in one or more offshore utilities.

The impact that the National Energy Policy Act had on California's energy market is quite significant, as it put added pressure on the state to open its market to competition.

CALIFORNIA'S UNPRECEDENTED MOVE TOWARD ELECTRIC POWER DEREGULATION

Without question, the National Energy Policy Act of 1992 can be seen as the fundamental catalyst that put California's move toward direct access into high gear. The 20-month period between April 1993 and December 1995 may well be remembered as an historic period in California, as least from the perspective of the state's diverse electric service providers and vocal consumers, large and small. At a time when electricity rates of the incumbent electric power utilities were 30–50% higher than the national average, large industrial customers developed a very influential lobbying force that in large part drove the efforts to deregulate California's electric power market. Large customers such as steel makers, mining concerns, and cement makers, for instance, recorded that their electricity costs made up 25% of their overhead.

Consequently, large industrial customers with headquarters in California that were subject to the state's high electricity rates were very vocal about wanting a break in their energy bills and argued that electric power competition would lower rates. For their part, the three incumbent IOUs wanted to be released from intense regulatory oversight from the CPUC and thus were also eager to see deregulation proceed. Consumer groups, the third stakeholder contingent in the debate of California deregulation, were fairly neutral on the concept of deregulation, while remaining steadfast in their opposition to any program that overly favored the huge utility companies.

Facing these dynamics, during this two-and-a-half-year time period, regulators and stakeholders hammered out a new competitive structure for the state's electric power industry, dismantling decades of rules, regulations, and preconceptions about how electric power should be bought and sold. The result was a competitive marketplace that tested theories about true competition, market dominance, and equal access.

In 1993, representatives from the CPUC traveled to the United Kingdom to study the electric power privatization model that had been developed there. In February 1993, the CPUC issued a policy statement, *California's Electric Services Industry: Perspectives on the Past, Strategies for the Future*, commonly referred to as the "Yellow Book." The Yellow Book provided an early, but still quite comprehensive, review of the regulatory conditions and future trends facing the electric services industry in California. Most significantly, the Yellow Book revealed the state's intention to construct an electric power competition model that would include an auction or power pool entity, as had been included in the U.K. model. The Yellow Book triggered extensive public comment and a full-panel hearing process that culminated in the subsequent policy report on electric power restructuring issued by the CPUC.

Moving an important step forward, on April 20, 1994, the CPUC made public a surprising proposal to restructure California's electric utilities, with the intention of making them more competitive. The lengthy proposal subsequently became known as the "Blue Book." The Blue Book proposed opening portions of California's electric power industry to further competition and introduced performance-based ratemaking (PBR) for the remaining elements of the industry. Under the new policy of PBR, incumbent utilities would be rewarded for reducing costs by setting rate adjustments, with formulas based on productivity factors and inflation indices. The electric power market would be divided into two sectors: the direct access sector, in which customers could buy electric power from their supplier of choice, and the utility service sector, in which incumbent utilities would still be required to deliver electric power to customers.

In essence, the CPUC's Blue Book set guiding principles for deregulation in California. Fundamental to the early model for competition were these:

- The preservation of the incumbent utilities' reasonable opportunity to earn a fair rate of return;
- The continuation of public purpose programs; and
- The continuation of safe, reliable, reasonably priced, and environmentally sensitive electric service.

Also key to the Blue Book proposals was the idea of vertical disintegration, or the separation of utility generation from transmission and distribution. The CPUC asserted that separating generation from transmission and distribution would add value to the restructuring of California's electric power market, allowing a "win-win" situation for utility ratepayers and shareholders. The added economic value would come from tapping potential cost savings in existing generation assets—by lowering the cost of capital, fuel costs, increasing plant availability, etc. By selling off generation assets, both utility shareholders and ratepayers would theoretically get an immediate benefit from subsequent restructuring.

In December 1995, the CPUC issued a final proposal to phase in direct access for electricity, beginning January 1, 1998 for customers taking service at the transmission level and for all customers by 2002. (*Note that the phase-in approach would be significantly changed by AB 1890, discussed next*). This decision recommended an industry structure that included a local distribution function, while allowing retail customers to enter into a wide variety of contractual arrangements with generation suppliers and other market participants. These contracts would have the commercial and operational flexibility desired by direct access advocates, while preserving state jurisdiction over local distribution. The policy also addressed market power and the need for a level playing field. In other words, the CPUC aimed to develop policies that would ensure that new competing generators would be able to enter the state and compete with incumbent utilities.

Main elements of the CPUC's ruling to facilitate the transition to competitive market included the following elements.

The transition process would begin on January 1, 1998, and be completed by 2003. During the first year long phase of restructuring, a selection of individual residential, commercial, agricultural, and industrial customers would be permitted to handle the following:

- Negotiate and sign contracts with power generation sources in a direct-access arrangement;
- Aggregate a number of individual loads through a power marketer or energy broker; or
- Continue purchasing electricity from their incumbent utility.

Creation of the Independent System Operator (ISO). The California ISO, which was still in operation January 2002 even though direct access had been terminated in the state, was originally referred to PoolCo and modeled in large part on the restructuring template used in the United Kingdom. In essence, California ISO, as established by the Blue Book, was envisioned as the main vehicle to ensure that competing bulk-power producers active in the state had equal opportunities to deliver their products to consumers. It was during the time of the Blue Book that the CPUC decided that the state's incumbent utilities should be required to transfer operational control of their transmission assets to the California ISO. In other words, the utilities would continue to own their transmission assets, but the California ISO would be granted operational control of the assets. Even in this early stage of development, the California ISO was envisioned as somewhat of a "traffic cop," to the extent that it would manage and coordinate activity throughout the transmission grid in the state.

The California ISO (or PoolCo as it was referred to in early stages of its development) would perform the following functions:

- Coordinate, integrate, and balance electric power production and consumption;
- Manage economic dispatch;
- Ensure system reliability; and
- Provide comparable transmission pricing and open access to the California grid system.

Creation of the Power Exchange (PX). During its 1995 ruling, the CPUC also decided that an independent power exchange (PX) would be created. The California PX, which was officially shut down January 2001, acted as an auction and essentially matched bids between buyers and sellers and then submitted a delivery schedule to the California ISO. Under the CPUC ruling, utilities under the CPUC's jurisdiction (essentially the three incumbent IOUs) would be required to buy and sell generation capacity from the PX. The only exception to the mandatory buy/sell rule was for generating facilities that had been divested by electric utilities. All California utilities were required to purchase electricity from the PX. The CPUC's justification for this mandatory-buy requirement was that even California customers who chose to remain with their incumbent utility would theoretically benefit from price reductions as they were anticipated to occur in the PX. Participation in the California PX would be voluntary for all buyers and sellers other than the IOUs, including the state's municipal utilities that had the option of participating in electric competition. However, remember that the IOUs served 70% of the California market; so consequently, any electric power generator or marketer that wanted to develop a substantial business in California was forced *de facto* to sell power to the California PX.

The California PX was developed to determine the hourly price of electricity. Prices were based upon supply and demand. Essentially, the California PX performed as a competitive spot market from which the California utilities purchased their electricity and to which they sold any minimal power that they still produced on their own. The California PX solicited bids from electricity buyers and generators and chose the lowest generation bidders until the

California PX had enough electricity supply to meet requests for power. The California PX's prices changed on an hourly basis.

Assembly Bill 1890

The first piece of legislation intended to realize the CPUC's goals was Assembly Bill 1890, also known as *The Electric Industry Restructuring Act.* AB 1890 was passed by the California Legislature without a dissenting vote in September 1996 and signed by then-Gov. Pete Wilson. First and foremost, AB 1890 is significant because it changed the original Blue Book implementation dates, making direct access available to all California customers on January 1, 1998 (this date would later be postponed to March 31, 1998, due to computer problems). In addition, AB 1890 ruled that electric power generators in California would be open to competition, but transmission and distribution of power would still remain regulated. Distribution companies (essentially the state's three IOUs) were ordered to begin accepting requests from their customers to change power providers on November 1, 1997.

AB 1890 is also important because it promoted the establishment of a competitive electric power generation market in California, initiated direct access in the state, and formally adopted the California PX/ISO market structure that had been proposed by the CPUC. In addition, AB 1890 also authorized the recovery of stranded costs for utilities (those costs related to the construction of generation facilities previously accrued by the IOUs). This was accomplished through the use of state-backed bonds, and mandated open, non-discriminatory access to transmission and distribution services for new competitors that wanted to enter the state. AB 1890 also called for a 10% rate reduction for residential and other small customers that would remain in effect through March 31, 2001 (or earlier if any of the three incumbent utilities were able to recover their stranded costs before that date).

Generally speaking, AB 1890 gained broad support among California stakeholders because it accomplished the following:

- Provided for a rapid and orderly transition to a competitive market-place in power generation;
- Assured an open and transparent power marketplace serving both large and small customers;
- Provided substantial early rate relief for the state's residential and small commercial customers with an automatic 10% discount off electricity rates; and
- Provided utilities with a fair opportunity to recover costs they incurred in meeting their legal obligation to serve all customers under the state's regulatory system.

Under the restructuring law, the state's three IOUs agreed to continue operating their distribution facilities. Under direct access, if a customer of the incumbent utility decided to purchase power from an alternate supplier, the incumbent IOU would still be responsible for delivering that power through the incumbent's transmission and distribution system.

In addition, AB 1890 also established a competitive transition charge (CTC), which would provide an accelerated recovery system for the utilities' stranded costs. The CTC, which would vary by utility, would appear on a customer's bill as a separate line item, the proceeds from which would be used to pay the utility's stranded costs. However, recovery of utility costs was already built into the existing regulatory structure and was included in rates charged to all customers. If there were no transition to a competitive market, customers would continue to repay these costs to utilities through their normal electricity bills. Consequently, the CTC did not cause an increase in electricity rates from existing levels and should not be viewed as an additional cost. The CTC was determined by multiplying a CTC rate by electrical energy consumption.

Another key policy of AB 1890 was the freezing of electricity rates. Regulated IOU rates for agricultural, residential, industrial, and large commercial customers were frozen at their June 1996 levels until utilities recovered their stranded costs through the accumulation of the CTC proceeds. As of January 1, 1998, rates for residential and small commercial customers in California (defined as 20 kilowatts or less peak demand) were automatically reduced by 10% and would remain at that level until the incumbent utility recovered its stranded costs or until March 31, 2002, whichever came first.

The automatic 10% rate freeze was often cited by pro-deregulation advocates as a key benefit of California's direct access program in the early days of electric power competition in the state. However, the very same rate freeze would also be one of the very first indicators of the flawed market in California. Specifically, California's vulnerability to volatile wholesale prices would be demonstrated when San Diego Gas & Electric paid off its stranded costs through the CTC, thus exposing its customers to skyrocketing costs for power in the summer of 2000.

Moreover, from the utilities' perspective, AB 1890 was very important because it granted 100% recovery of stranded costs to the three IOUs in California, a decision that was revolutionary. The decision to grant 100% stranded cost recovery was contingent upon the IOUs guaranteeing to a rate freeze that would theoretically protect customers from price fluctuations. In other words, the IOUs realized that the only way to secure 100% stranded cost recovery was to concede to a rate freeze for four years. The utilities widely supported this bill because the benefits to them outweighed the costs.

During 1996 to 1998, the CPUC finalized its plans for initiating direct access and implementing the tenets of AB 1890. As noted, the CPUC worked toward a start-date of January 1, 1998, to launch direct access for all California customers. The CPUC also approved a rate-reduction bond financing that provided the incumbent utilities with up-front funding for a portion of their stranded costs.

At this point, the incumbent utilities and other new developers were even more hesitant to invest any of their own capital in new generation plants, at least until they saw how the deregulation law would be implemented in California. Per negotiations with the CPUC, the incumbent utilities began a process of divesting their own generation assets to independent, unregulated companies. A total of 18,393 MW of generating capacity exchanged hands during this time, changing ownership from the three California utilities to out-of-state companies that would in turn sell electric power into the auction system of the California PX.

Federal Measures Also Drove California Deregulation

While the various state measures were being developed in California, federal policy also was advancing the national movement toward electric power restructuring. In 1996, FERC issued Order 888, a significant order that stated public utilities should open their power lines to any electric generating company willing to pay a fair transmission cost. Order 888 created an environment for the wholesale trading of electricity (between generators and customers regardless of their location in the United States), which in turn helped California to implement competition on the retail level. The Order also forced utilities to separate their transmission businesses from their power-marketing units, and paved the way for utility divestiture of power plants. Most importantly, Order 888 can be seen as creating a new market for wholesale electricity that has enabled utilities and other marketers to buy and sell power at unregulated prices. Ongoing ramifications related to Order 888 include concerns from FERC commissioners and some industry officials, who say that discrimination continues to take place by utilities that are favoring the wholesale trading activities of their own affiliates by giving them preferential access to the transmission lines that they operate.

Moreover, after six years of planning and intense debate, the state of California was set for electric power competition to begin in California in the spring of 1998. The plan seemed fairly practical and almost simple in its objective. Utilities would be required to buy power on the open market through the newly established PX, which in theory would bring lower electricity rates to California customers. In addition, the rate freeze, which would remain in effect until March 2002, would protect California customers from any fluctuations that might occur on the wholesale market. Utilities would have an opportunity from 1998 until 2002 to sell their power-generating assets and recover their stranded costs. The fact that the utilities would have little generating assets of their own was thought to be a foolproof catalyst for jumpstarting electric power competition in the state.

The New Frontier

To say that the California's electric power restructuring plan created a new market in the state would clearly be an understatement. It is no stretch to call the energy market in California created by AB 1890 a "new frontier." In fact, the California energy market created by the array of stakeholders and implemented by AB 1890 would have little resemblance to the market that preceded it.

Here is how the new market in California looked as of April 1998, and how it would remain until January 2001, when the State of California began to re-regulate the state's electricity market. Under the new restructuring law created by AB 1890, the transmission and distribution of California's electricity market would remain a regulated monopoly, but the generation of electricity in the state would be opened to competition. The three IOUs impacted by this restructuring law—PG&E Co., SCE, and SDG&E—were strongly encouraged to sell their power plants and were also required to purchase all of their electricity from the newly-formed auction system on the wholesale market.

As of March 31, 1998, all customers located in the service territories of California's three IOUs were allowed to shop for electric power in an open market. In other words, these customers were no longer restricted to buying power solely from their incumbent utility. Direct access gave Californians the ability to shop around for energy deals and select the one that best met their needs. For most of these customers, AB 1890 froze electricity rates at a level 10% below their 1996 level. The frozen rates were expected to be above the utilities' actual costs, thus giving the utilities a chance to recover their past generation investments, while guaranteeing lower rates for consumers. The rate freeze was to last until the utilities recovered their past investments or until March 31, 2002, whichever came first. Once the freeze was lifted for each of the three utilities, electricity costs would fluctuate with the market.

The California ISO assumed its new role of maintaining overall electric system reliability in the state. The California PX assumed its role of creating a "pool" or "spot market" in which price information about electricity sales would become publicly available.

PG&E Co., SCE, and SDG&E continued to own their own transmission facilities, but turned operational control of these facilities over to the California ISO. The California ISO, acting like a air traffic controller or "traffic cop," assumed operational control over the state's transmission system to ensure that electricity flowing into California reached all customers, so they would continue to have reliable service when they needed it. The California ISO was also charged with making sure that all generators had equal opportunity to send their electricity through the transmission system to their customers.

The incumbent utilities continued to operate their distribution lines and remained responsible for reliable, safe delivery of electricity. The CPUC assumed the role of overseeing the three state IOUs to make sure that they fulfilled this obligation. The CPUC also became in charge of regulating the transmission and distribution rates of the utilities, using performance-based, rather than cost-of-service ratemaking.

The three IOUs were required to sell any electric power that they still had to the PX. Municipalities, independent power producers, irrigation districts, and out-of-state producers also essentially were forced to sell power to the PX, as the incumbent utilities had no choice but to buy power from the PX. In other words, if these power generators wanted to sell power to the incumbent utilities, which represented 70% of the California market, they would have to sell their power in the California PX. The IOUs paid a price determined by the PX based on market demand for power. In theory, this was thought to ensure fair competition between utilities and other electricity suppliers.

Whereas, once generating plants were owned by incumbent utilities that set prices regulated by the CPUC, now electric prices in California would be set at auction through the California PX. Further, the utilities were required to buy and sell power through the PX, and were prohibited from purchasing power from sources outside the PX. The utilities had to buy power on the PX's spot market, because the PX did not offer power at guaranteed prices. Whereas, once utilities that generated electricity also owned transmission lines and the power grid, control of those assets would now be transferred to the California ISO.

The only similarity between the previous market in California and the new frontier would be that utilities would continue to own and control the state's distribution system (the wires that supply California homes and businesses with power). The expectation for the newly created market was that electric power competition in California would drive down the cost of electricity, leading to lower rates for everyone.

That, at least, was the plan. Little did state legislators, the CPUC, the three IOUs, or unsuspecting end-user customers know that with this new restructuring plan, the state of California was headed toward a major meltdown. The concept of electric power deregulation that had been seemingly well planned would amount to a recipe for disaster in little more than a year, with failure of California's energy market having a huge negative impact on the entire state and possibly the rest of the nation.

Chapter 2
The Root Causes of the California Energy Crisis

It might have been seen as a bad omen. California's widely publicized retail choice program did not begin on its scheduled date of January 1, 1998, as outlined by AB 1890. The start date was extended to March 31, 1998, as a result of the variety of computer and planning problems faced by the governing boards of the California ISO and the California PX. Officially, the CPUC said it delayed the start date of retail competition because additional time was needed to test computer software at the California ISO and PX. Despite the delay, the 10% rate reduction for all California residential and small commercial customers mandated by AB 1890 went into effect January 1, 1998.

The delay in the start of direct access in California was not necessarily a surprise, considering the massive undertaking that deregulating the state's electric power market became. California's power grid, a transmission system made up of high-voltage power lines supported by 100- to 150-foot towers, delivers 164-billion kilowatt-hours (kWh) of electricity each year, enough power to serve the annual energy needs of the 27 million customers of the state's three IOUs.

However, despite this rather inauspicious beginning, the first two years of California's direct access program ran fairly smoothly, at least from an operational standpoint and when compared to the problems that lay ahead. Although the operational foundation of the state's competitive electric power market ran fairly smoothly in the first year of direct access, this did not translate into a large number of Californians switching to a new electric power provider. As of May 1, 1999, just over a year after direct access had started in California, only 3–4% of 70% served by the incumbent IOUs were participating in the choice programs available through the local distribution company service areas. The participation at that level represented a decline from the initial 5% participation that had been measured at the beginning of direct access in April 1998.

Thus, direct access in California was never what could be considered a great success. Despite the huge sums of money that the incumbent utilities spent on educating their customers about electric power supplier choice, for the most part Californians remained apathetic about selecting a new power supplier and opted to stay with their incumbent utility. According to data available from the CEC, even by the end of 2000, only 1% of residential customers in California had switched electricity providers, while between 15–20% of industrial customers—the class that had advocated direct access in the first place—had opted to find a new provider.

Keep in mind that, despite these market changes, restructuring laws and regulatory policies in the state of California remained intact. For instance, under the state's rules, the three IOUs were required to buy all of their power through the California PX. The utilities were not permitted to enter into forward long-term contracts for energy. The California PX, as a spot-market auction pool, reflected the fluctuating cost of electric power. As a result, the California IOUs were also subjected to the dramatic increase in the price of wholesale electric power, which would increasingly create a situation of enormous debts for all three utilities.

REASONS FOR THE ENERGY CRISIS

There are many different explanations as to why the California energy crisis occurred. In fact, as the situation worsened during the time period of 1999 to early 2001, all of the various involved parties developed their own interpretations as to the causes for the crisis, and identified who they thought was to blame.

Nevertheless, despite an ever increasing round of finger pointing and innumerable theories on fundamental causes, the best general consensus available is that the California energy crisis resulted from a volatile combination of the following eight factors:

1. California's wholesale electricity market was inherently flawed.
2. Power supply did not keep pace with demand.
3. Power imports from other states decreased.
4. A large percentage of in-state generation was off line.
5. Wholesale prices for electricity began to rise.
6. The transmission grid in California became dangerously strained.
7. Weather patterns took a toll on the California energy market.
8. In addition, in one of the most contentious debates over the California energy crisis, California officials repeatedly have blamed out-of-state generators for "gaming the system" and manipulating California's deregulated market.

Let's look at each of these causes individually.

California's Wholesale Electricity Market Was Inherently Flawed

Since the incumbent utilities in California were barred from engaging in bilateral contracts for their electric power supply and forced to buy power from the California PX, they were at the mercy of the wide fluctuations of the wholesale market with no available recourse. The allowance of bilateral contracts would have allowed the utilities to minimize risks and stabilize prices. However, per the requirements of AB 1890, the California IOUs were required to purchase all of their energy on the spot market, where prices fluctuated. Second, energy pricing in the California PX was set so that the most expensive unit of energy would determine the price of all units. Third, because Californians were protected by rate freeze and thus not exposed to the true price of electricity, they had no incentive to conserve or reduce their demand. Consequently, demand continued to rise, and the utilities were caught in the vicious cycle of maintaining their obligation to serve and being required to buy electric power from the California PX.

The CPUC pressured the three California IOUs—PG&E Co, SCE, and SDG&E—to sell off their generation plants in an attempt to limit any potential market power that the three utilities would have. In addition, the California model included a PX from which the IOUs would be required to buy the power they needed to serve customers on a day-ahead, spot market. The IOUs also had to sell into the California PX any power they continued to own or produce. The law also gave residential customers an immediate 10% rate cut based on 1996 rates.

All of this amounts to what is perhaps the fatal flaw of California's market. The three IOUs were encouraged to sell off their own generation assets, yet were given no guaranteed access to affordable electric power. The utilities were forced to buy power from the PX, where prices skyrocketed beginning in the summer of 1999 due to short supplies and increased demand. At the same time, a rate freeze kept all three utilities from charging customers a rate that would allow for direct compensation of the price of wholesale power.

By stripping the California IOUs of the ability to manage their generation supply, the California restructuring law dismantled the integrated utility model in the state. As a result, more or less, PG&E Co. and SCE found themselves over $12 billion in debt, causing PG&E Co. to declare bankruptcy, SDG&E customers paid bills that were double and triple the 1999 levels, and deregulation in the state appeared to be a complete failure. As a result, the movement toward re-regulation in California appears to be gaining momentum.

The only utility in California that seems to have had any success since deregulation began in the state is the Los Angeles Department of Water and Power (LADWP), a municipal utility that opted out of deregulation and maintained the integrated utility model by keeping its generation assets. Early on, S. David Freeman, LADWP's general manager, made a critical decision—one that in hindsight appears to be flawless—for LADWP to opt out of deregulation while the state's three IOUs continued down the treacherous path of restructuring. While the IOUs began divesting themselves of generating facilities as a means of averting market power issues (due to agreements established with the CPUC), LADWP was a prime example of public power throughout the United States electing to keep the integrated utility model. In addition, Freeman blocked ongoing attempts by a consortium led by Duke Energy to buy LADWP's trading unit, and thus maintained control over the municipal's generation assets.

With demand on an unexpected increase and supplies in questionable availability, LADWP found itself having the upper hand. The wholesale market clearly favored generators, and LADWP was one of an elite group of companies that capitalized on the disparity between supply and demand. Critics accused LADWP of "profiteering," but Freeman contended that the utility was just taking advantage of California's flawed market. The truth is that LADWP was in the right place at the right time, and maximized its ability to generate cash via power trading, which it used to drive down its debt.

And, while the state IOUs are presently dependent on the California state government to supply them with wholesale power, LADWP is free to continue with an aggressive generation plan. Among its objectives is to add 2,900 MW of power generation within its service territory in the Los Angeles basin. The end result of this would be less dependence on external sources for electric power and instant supply available at times of peak demand.

Power Supply Did Not Keep Pace with Demand

There is little debate over the claim that demand in California rose to unprecedented levels in the mid-1990s. About 600,000 new residents move to California every year, and add to the population of what is already the nation's largest state. According to information from state officials, many of the new residents of California move to the Central Valley region, where temperatures are extremely hot in the summer and air conditioners are used daily. In addition, soaring use of computers and other devices used to run the so-called "New Economy," along with the increased need of natural gas to run in-state power plants, contributed significantly to the increase of demand in California. Production figures from the Edison Electric Institute, the national group that represents the interests of IOUs, reported that consumption in California grew by 4% in 1996, 3.4% in 1997, nearly 5% in 1999, and a whopping 10% in 2000.

Despite this increase in demand, investment in new power plant construction in California dropped off significantly in the years before direct access actually began in the spring of 1998. As would be often repeated in the debate over the impending energy crisis, no new large power plant had been built in California in over a decade. Per negotiations with the CPUC, California's utilities were encouraged to sell their power plants and were not responsible for building new ones. In addition to the divestiture, utilities that had previously owned and controlled their own generation supply in California did not have much incentive to build new power plants in the years before direct access started. First, environmental concerns—which traditionally have been a major focus of public debate in California—were seen by many as an immediate roadblock to any plans that might have been pursued for new generation. Specifically, state regulators at the CPUC placed more and more restrictions on power plant construction, until any plans for new generation in the state were effectively halted.

Further, for at least two years before AB 1890 was passed, the California utilities held off building new power plants because they didn't want to spend millions to add new capacity when they knew that deregulation was imminent, a process which would most likely include the utilities selling off generation assets. In other words, given the uncertainty of how deregulation might impact the state's electricity market, the incumbent utilities and other energy companies naturally were not inclined to make significant investments in new power plants, despite the growing concern about a supply/demand imbalance. Clearly, the power supply in California would not be capable of keeping pace with demand, which was on the rise. Specifically, California's generation capability decreased 2% from 1990 through 1999, at the very time when retail sales in the state increased by 11%.

Ironically, when the California Assembly approved AB 1890 in 1996 and even until the time when direct access began on March 31, 1998, the state of California appeared to have an adequate supply of power. However, a disastrous combination of increased demand, aging power plants, and the ability to import less power from other states collided to create service reliability problems in the state. By the early part of the summer of 2000, the North American Electric Reliability Council (NERC) issued a warning that while California would possibly have an adequate supply of power during the summer season given normal weather patterns, the region might not have "adequate resources to accommodate a widespread severe heat wave or higher than normal generator outages." The NERC warnings proved prophetic when the summer of 2000 turned out to be an unusually hot season throughout California.

The ensuing crisis would result in California finding its reserve capacity, the safety margin of extra power available for emergencies, at its lowest level in more than three decades. Specifically, at some points in the summer of 2000, power use hit record levels of 46,245 MW, when officials had established reserves of 46,400 MW, leaving a very slim margin for error.

Power Imports from Other States Decreased

The vast majority of California's natural gas supply—about 60%—comes from other states, mainly along the Rocky Mountains, and about 25% comes from Canada. In addition, California has traditionally relied on imports of hydroelectric power from the Northwest to supplement its internal state supply of power. California and the Northwest have historically engaged in a mutual power trading partnership. California normally trades or sells power at cheaper rates to the Northwest in the wintertime, and the Northwest in turn trades or sells power back to California at cheaper rates in the summer. The amount of power being exchanged typically constitutes 20% of the load for the two regions. That partnership worked well until the summer of 2000, when short supplies, increased demand, and unusually hot temperatures began wreaking havoc on the California market and reduced the amount flowing from the Northwest.

Reports vary, but as an estimate, California has traditionally imported between 7 and 11 gigawatts of out-of-state generation capability. However, during the early summer of 2000, the Pacific Northwest witnessed the beginning of what would become its worst drought in history. An unprecedented drop in the amount of rain and snow severely lowered its own hydroelectric supplies and, consequently, the amount of power that could be exported to California. In addition, growth in the Pacific Northwest has been dramatic, which also reduced the amount of energy that was available to export to other states such as California. Other areas of the country also faced tightening supplies, which also limited the amount of power that California could import.

The fact that California's own power supply had not kept pace with the state's demand for power is certainly interrelated to the decreasing power supplies available from neighboring states. For years prior to the start of direct access in California, there had been no new construction of power plants in the state. Thus, as California's demand increased, the state became heavily dependent on power imports from the Pacific Northwest and the Southwest. When demand in those regions also increased, there was less power available to send to California, which exacerbated California's already significant power supply problems. For instance, the Northwest's demand for electricity reportedly grew at a steady rate in the 1990s, due in large part to the demands of the high-tech industry that is so prominent in that region. The impact on California was significant, because when the Golden State looked north for supplemental power, there was less to be shared.

Further, while the California crisis continued to worsen, the Pacific Northwest was facing its own growing challenges, many of which had resulted from its southern neighbor's problems. In addition to this domino effect, the Northwest began encountering similar problems all its own. Just as California faced an unusually hot summer, Oregon and Washington faced their own weather-related problem—the lack of rain. A long dry spell severely diminished the productivity of the hydroelectric dams in the Pacific Northwest, shrinking what was commonly a back-up form of power during the winter months. Reportedly, reservoir levels during 2000 were half of normal, and stream flows into them were only 30% of normal. As a result, utilities that had depended on hydroelectric power were forced to supplement their energy needs with power purchases on the wholesale market.

Also, like California, supply did not keep pace with demand in the Pacific Northwest. While the population has increased dramatically in the area, most reports indicate that construction of new power plants has not been a high priority for regulators in the two Northwest states. The reasons for the lack of new power plants include environmental concerns, permitting restrictions, and high costs associated with the plant construction. In addition, just as in California, some power plants in the Northwest have been closed recently for maintenance, which also exacerbated the concerns about where power supplies would be generated. Compounding the situation, an unusually cold winter of 2000-2001 moved into the region, and as a result demand rose steadily for the next several months.

This problem continues today. Despite the fact that new plants should come online between 2002 and 2004, presently California must import about 11,260 MW of power (in addition to the 45,565 MW it generates within the state). Due to the tightness of the market, wholesalers have been charging high electricity rates, which were passed onto SDG&E customers until an emergency rate freeze was put into place. PG&E Co. and SCE customers were still protected by a rate freeze, however the utilities themselves were exposed to the high costs of power that needed to be imported from the wholesale market in other states.

A Large Percentage of In-state Generation Was Off-Line

Over the course of 2000, some 10 gigawatts of generation in California was taken off-line for routine maintenance upgrades or as a result of system problems. During the winter of 2000-2001 and the spring of 2001, average daily outages ranged from 9,000 to 15,000 MW out of a potential capacity of 42,000 MW in the California ISO territory. Some plants across the state were involuntarily taken off line because of expired nitrogen oxide emission credits. The problem became particularly acute toward the end of 2000, which unfortunately collided with unfavorable weather patterns that were independently driving up demand. At some points during the course of the California energy crisis, approximately 25% of the state's generation pool was unavailable. That is about three times the amount of electricity generating capacity that was off-line under normal circumstances in California.

It is important to note that unavailability of power in California was due to both scheduled and unexpected factors. For instance, in December 2000, the 1,100-MW Diablo Canyon 2 nuclear power plant owned by PG&E Co. went off-line as part of scheduled maintenance to clean its water intake system. This was something that officials monitoring the available power supply in California could plan for, even though it heightened uncertainty about the vulnerable supply/demand balance. At the same time, some unexpected outages also occurred. Reportedly, about 13,400 MW of power were lost due to a transformer fire at a Southern California plant and other idled plants across the state during the latter part of 2000.

In addition, about 2,000 to 3,000 MW of production from QFs were lost at one time when the state experienced rolling blackouts. The QFs that were off-line typically had produced only small amounts of power under contract to the state's three major utilities. However, the QFs had stopped producing power at certain times in 2000 as a result of the financial problems of PG&E and SCE and the fact that they had been paid only a small percentage of what they were owed for power previously produced. Under normal circumstances, the QFs can generate as much as 30% of the state's electricity needs.

According to reports from the California ISO, an average of about 11,000 MW was scheduled to be off-line for most of December 2000. Of the 11,000 MW that were off-line during that month, 7,000 MW were off-line for unplanned maintenance, including those plants taken down for expired air emission credits. For the year 2001, statewide in California, 16 units that could generate a combined 5,225 MW—more than 10% of the state's power output—were scheduled to be shut down at some point during the year, mostly for the installation of selective catalytic reduction equipment that limits nitrogen oxide (NO_x).

It is important to note that, once generation is taken off-line, it is often difficult to power up a particular plant within a short period of time. Thus, generation that is off-line cannot be quickly started if needed to meet a sudden increase in demand.

The fact that a large percentage of in-state generation was off-line in California sparked intense investigation among state regulators. Remember that, due to negotiations between the state IOUs and the CPUC, the IOUs had divested the vast majority of their power generation assets. The assets had been bought primarily by out-of-state generating companies, which in turn sold the output from the plants back into the California PX. In September 2000, CPUC Commissioner Carl Wood expressed concern that several generation units might have been taken off-line due to reasons that may have not been legitimate. Moreover, the CPUC launched an investigation of how electricity was bought and sold during specific time periods in order to determine that any power plant that had been taken off line had a legitimate reason to do so. Part of the ambiguity related to this issue is that, under a deregulated environment, the California ISO does not publicly disclose information about which units might be coming off-line because such information could drive up power costs.

Some California officials suggested that the unscheduled outages directly caused wholesale prices in the state to be more than 10 times higher than they had been in 1999. However, in response, energy companies that owned the plants argued that in general the plants had been working especially hard during the summer of 2000, during which time California experienced warmer-than-normal temperatures. The overworking of the plants caused more maintenance than typically would be needed.

In February 2001, FERC concluded that it lacked proof to conclude that power plant shutdowns in California during the winter of 2000-2001 were instigated to drive up electricity prices in California. FERC said that telephone interviews, a review of four years of documentation, and on-site inspections did not yield any evidence that producers "were scheduling maintenance or incurring outages in an effort to influence prices." Instead, the audit found that "the companies appeared to have taken whatever steps were necessary to bring the generating facilities back online as soon as possible by accelerating maintenance and incurring additional expenses." For the current probe, federal regulators visited three power plants in the Los Angeles area—two owned by Reliant Energy and one owned by a joint venture of Dynegy and NRG. The plants were selected randomly, based on the size of facilities that went down for repairs in December and their geographic proximity.

Federal investigators found that the plants that went off-line were 30 to 40 years old and had been running at a significantly higher rate than in previous years. "Most of the generating facilities were out of service because of tube leaks and casing problems, turbine seal leaks and turbine blade wear, valve failure, pump and pump motor failures," the report found. The audit's findings drew sharp skepticism from both consumer advocates and a spokesman for Gov. Gray Davis, who had repeatedly criticized the staunchly free-market commission for not placing price caps on the western electricity market. However, the audit did not look into separate allegations that power firms were withholding electricity from the market by refusing to sell or by pricing power so high that buyers would not make an offer.

Wholesale Prices for Electricity Began to Rise

Faced with a natural gas shortage, soaring oil prices, and unprecedented demands for electricity, the United States encountered a severe energy crunch during the winter of 2000-2001. As a whole, America's supply of natural gas had also been declining since the mid-1990s, when energy companies began cutting back on production as prices fell. However, most power plants in California (and many other states, for that matter) are powered by natural gas, and during the time period of 1999 to early 2001, the price of natural gas on the wholesale market embarked on a dramatic increase. As a result of an increase in natural gas prices, increasing demand, and the high cost of meeting California's power plant emissions standards, the cost of wholesale power—which the California utilities were locked into buying from the California PX due to AB 1890—began to reach sky-high levels. Even absent the dramatic shortage that was obvious in the state of California, the cost of making electricity increased dramatically from 1999 to early 2001. Most power plants in California, which are now owned by out-of-state companies, are powered by natural gas, and the rise in natural-gas prices caused electricity prices to increase dramatically as well.

Further, pollution control credits, which some power plants must buy to operate, dramatically increased in cost, which again contributed to the price increase. As a result of these factors, starting in June 2000, the price for energy skyrocketed far in excess of what the utility companies were allowed to recover from their customers in rates. PG&E Co., for example, was reportedly required to spend more than five times as much to purchase power as it was allowed to charge its customers. The result of this inequity would be a mounting under-collection that by the end of 2000 would leave the utility in debts totaling $6.6 billion (a figure that would continue to increase in 2001).

Here are some specific figures that can be used to illustrate the high costs of wholesale electric power during 2000 and early 2001:

- In September 2000, electric power traded between $175 to $179/MWh.
- In early November 2000, natural gas prices in California hit record levels, trading at nearly $19/MMBtu due to pipeline problems and strong demand. At that time, natural-gas prices in California were reportedly five times higher than at the same time in the previous year.
- In early December 2000, electricity forward prices in the Mid-Columbia River hub were higher than they were in the summer of 2000, when demand outstripped supply in California. In early December 2000, January future prices were trading at $474/MWh, and balance-December was trading at $525/MWh. Third-quarter trades were at $250/MWh.
- Also in early December 2000, natural gas for January 2001 delivery rose 76 cents, or 11%, to $7.43/MMBtu on the New York Stock Exchange. Earlier in the session, trading was halted when the price of natural gas jumped 19% to $7.95/MMBtu, the highest amount and the biggest one-day gain in 10 years of trading on the New York exchange.
- New York Mercantile Exchange gas prices for December 2000 through March 2001 reached record levels, boosted by weather forecasts calling for cold temperatures after the Thanksgiving 2000 weekend.

The Transmission Grid in California Became Dangerously Strained

In addition to constraints on generation supply, California's energy market was also impacted by a severely strained transmission system. Path 15, a narrow 90-mile stretch of power lines near Los Banos, CA, that carries electricity between Northern and Southern California, generally has less transmission capacity than other portions of the power grid. Path 15 became congested at times during the course of the summer of 2000. This congestion reduced the flow of surplus energy capacity in southern California to meet shortages in northern California.

The transmission congestion led to real-world problems during 2000 and 2001. For instance, in January 2001, Northern California suffered power outages in part because Path 15 could not deliver the surplus electricity that was available to Southern California. In addition, the California ISO estimated that congestion generated by Path 15 added $221.7 million to California's power bill in a 16-month period that ended in December 2000.

California's transmission problems were emblematic of the insufficient transmission systems in the United States. Usually when California's resources become low, the state can import energy from other states. But under the extreme circumstances of a heat wave, California—and many other states that want to import power—found that the transmission system had become clogged and power could not be delivered to all of the states that needed it. Thus, California found itself in a desperate situation in which its own power reserves were running low and there was little hope of importing power from elsewhere.

As will be discussed in greater detail in chapter 5, PG&E Co, which owns the existing Path 15, has joined with a group of other energy companies to spend about $300 million on an upgrade plan that will add a new line to Path 15. While this upgrade is necessary and important, most likely it will not be completed until 2002 or 2003 at the earliest, which means that one of the fundamental causes of the California energy crisis will remain intact for the near term.

Weather Patterns Took a Toll on the California Energy Market

Weather patterns during the summer and winter of 2000 also had a direct impact on the California energy crisis, and clearly contributed to the soaring demand (and resulting high prices) for electricity in the state.

The most recent problem appears to have resulted from unusually warm weather in the southern part of the state that drove up use of electricity for air conditioning. This left the state as a whole with limited reserves at a time when power plant maintenance and a lack of hydropower already was causing problems in Northern California.

The summer of 2000 was unusually warm for most of California, which unfortunately exacerbated the problem of power supplies that had already become severely compromised. According to the National Weather Service, heat in California during the summer of 2000 was steadily above normal. Along with independent reports from the California ISO, the indication was that the state could not cover its power demand when the temperature climbed above normal levels, even with every generator running and transmission lines all working at full capacity.

This condition continued into the early fall. The National Oceanic and Atmospheric Administration reported that the Western United States faced a 33% chance of seeing above-normal temperatures in September 2000 and a 50% chance of seeing above-normal temperatures throughout September, October, and November of that year. For instance, in September 2000, temperatures set records in four Bay Area cities, hitting 102 degrees in Gilroy, 97 degrees in Redwood City, 94 degrees in Mountain View, and 92 degrees in Oakland. At the same time, temperatures also reached 103 at Barstow and China Lake, CA. As a result of the heat and related increased demand, the California ISO issued 31 Stage One and 17 Stage Two alerts during the summer of 2000, indicating that power reserves had fallen to dangerous levels.

During the winter of 2000-2001, the Pacific Northwest faced temperatures that plunged below zero degrees Fahrenheit, with full-day averages ranging from the single digits to the 20s. In fact, the cold Arctic air dipped south from Canada and covered most of the western United States.

The cold front drove temperatures down and drove up the demand for electricity throughout the West at a time when supplies were already short.

Thus, demand for power in California during the winter of 2000-2001 increased sharply and continued to compromise supplies that had already been impacted due to the previous summer. In addition, the unusually cold weather in the Northwest that winter caused a spike in power demand in areas with a heavy reliance on electric heating and also contributed to a reduction in power that was available to export to California.

The hot summer and cold winter in California and the West also had the common denominator of being exceedingly dry. The dry conditions reduced the amount of water behind power-generating dams in the Northwest, which also impacted the California market because there was less power to export from the Northwest to California.

Market Manipulation by Power Generators

As noted, the debate over the extent to which out-of-state power generators that sold power to the three IOUs in California contributed to the ensuing power crisis became one of the most contentious war of words surrounding the state's energy market. Certainly, Gov. Gray Davis remained adamant in his belief that out-of-state generators had engaged in manipulation of the markets (at best) and outright price gouging (at worst) in the California wholesale markets. This situation resulted in the financial problems of the utilities and a growing amount of power costs that would ultimately be transferred onto California customers. Further, in March 2001, FERC moved to ensure just and reasonable rates for wholesale power sales in California by putting 13 California power sellers on notice that they must either make refunds for certain power sales or provide further justification of their prices, implying that the generators had overcharged for power sold to California.

As noted, power plant construction in California lagged while demand expanded. Perhaps leaders miscalculated how much demand would increase, and how much power would be available to meet California's power needs. In order to encourage generators to create as much power as possible, the

California law guaranteed the highest possible price for wholesale electricity. Through what became known as the "market clearing price," the last bidders, who are usually the most expensive, set the price that everyone would receive. The result was a market in which generators had clearly benefited financially. Reserve levels had fallen so dramatically in California that generators knew utilities would have to buy power from them, even as prices increased. In fact, power suppliers operating in the state found a financial windfall while utilities and end-user customers have seen their own bills increase exorbitantly.

However, the accusations of price manipulation or price gouging became a very serious matter and one that both sides of the argument defended fiercely. Further complicating the matter was the ambiguity over the term "price gouging" and how and when a company could be accused of engaging in this practice. U.S. antitrust law does not formally define price gouging. Under common understanding of the term, if a seller charges a price significantly higher than the cost of producing the product, including a reasonable return on the capital invested, then the seller has engaged in price gouging. At least this was the perspective espoused by Gov. Gray Davis and others who routinely placed blame at the feet of power generators active in California. The counter argument to this accusation is that sellers always set a price that will maximize profits, and price is always set by market and supply trends. Further, a company cannot survive if it does not set a profit-maximizing price, because investors will quickly sell shares of the stock of the company and purchase shares of other companies that are more focused on charging profit-maximizing prices.

Power generators active in California were also accused of holding market power in the state. The argument went that the state's IOUs, which had been stripped of their authority to manage their own generation supply, were forced to purchase power from the California PX, to which the generators sold their power. Knowing that they had a captive market, so goes the argument, power generators knew that they could charge high prices for wholesale power and the California IOUs would have no other choice but to purchase the power. Other accusations were that power generating companies that had purchased power plants from California IOUs had taken the plants off-line deliberately to spike prices, a claim that was never conclusively proven and one that generating companies repeatedly denied.

The unilateral exercise of market power is not illegal under U.S. antitrust law. However, the exercise of market power through coordinated action among various companies is illegal. Therefore, bidding extremely high prices or refusing to sell output in order to drive up the market price is not illegal under U.S. antitrust law as long as this behavior is not the result of coordinated actions by companies in the marketplace. Herein lies the legal ambiguity because, even though the unilateral exercise of market power is not illegal, electricity prices that result from the unilateral exercise of market power are illegal. Consequently, FERC became very absorbed in whether or not the electricity prices in California, which it deemed to be "unjust and unreasonable," were actually illegal and warranted refunds from the generators back to the state of California.

The accusations against the power generators were primarily focused on the following power generating companies, which to one extent or another were charged with taking advantage of the restructuring model in California:

- Arizona Public Service
- Avista Energy
- Duke Energy Trading and Marketing
- Dynegy Power Marketing
- Enron
- Mirant Corp.
- Nevada Power Co.
- Portland General Electric Co.
- Public Service Co. of Colorado
- Public Service Co. of New Mexico
- Reliant Energy
- Williams Energy Services

According to a report by Reuters, as of early 2001, of these power suppliers the four with the biggest potential refunds are Dynegy Power Marketing ($22.459 million), Duke Energy Trading and Marketing

($17.885 million), Reliant Energy Services ($12.435 million), and Williams Energy Services (amount owned not calculated).

The same companies responded with indignation to FERC's refund order. For instance, Duke Energy, Reliant, and Mirant responded that they would justify the prices previously charged rather than pay the refunds. Perhaps espousing a sentiment shared by most of the 13 power suppliers that are accused, a Mirant spokesperson said, "We don't feel we've done anything to warrant the refunds." Williams Energy Services noted, "it is a bit ironic that we might be asked to refund money that we haven't been paid yet," referring to the outstanding debts owed to the power suppliers by Pacific Gas & Electric Co. and SCE, California's debt-ridden utilities.

However, the extent to which these companies had clearly engaged in any wrongful or illegal behavior remains an unresolved topic even today. In December 2000, FERC ruled that California's deregulated market had resulted in "unjust and unreasonable rates" for the state's three utilities (Pacific Gas & Electric Co., SCE, and San Diego Gas & Electric). Nevertheless, the handful of power suppliers expected to pay the largest refunds have repeatedly maintained that they have not engaged in any price gouging or market manipulation with regard to their sales transactions in California's spot market.

The other indicator that was used to support accusations against the power generators was the strong increase in profits that these companies reported over the course of 2000 and into 2001. (See chapter 4 for detailed information on the profits reported by power generators active in the California market during 1999-2001.) Power companies rocketed past other U.S. industries in 2000, boosted by profits from soaring electricity prices that left California utilities nearly bankrupt and consumers bracing for higher bills. In a year that saw the Dow Jones industrial average fall nearly 5%, power companies reported returns to investors approaching 60%. Unregulated companies that now own a third of California's power generation, as well as out-of-state utilities and electricity traders, have reported gains of several hundred percent in 2000 earnings.

All of the companies, either overtly or subtly, have tried to dilute any direct correlation between their earnings and the high-priced California wholesale market. Dynegy's CEO made the following comment, which seems to be representative of the standard position among the other power suppliers: "Earnings from Dynegy's West Coast generation were not material and the company recorded what it considers to be an appropriate reserve for its California generation receivables."

Yet, most financial analysts have concluded that it would be ridiculous to claim that California has not boosted suppliers' bottom lines. Going much further, California Gov. Gray Davis overtly accused power generators of price gouging (or improperly inflating the amount they are charging utility buyers for power on the wholesale market) and gave the state attorney general $4 million to investigate and prosecute market-manipulation cases. Davis even proposed making it a criminal act punishable by a prison term for generating companies to withhold electricity.

However, proving which power suppliers scored the greatest financial profits in California is difficult because companies are not required to separate California from their overall earnings. FERC concluded in December 2000 that it found California wholesale prices to be "unjust and unreasonable," but could not confirm any wrongdoing. With that finding, power suppliers have wisely kept quiet about the extent to which California factored into their profits, especially considering that refunds might be required if any market manipulation is ever verified. The other reason to remain mum about California is that, due to the bankruptcy surrounding PG&E Co. and the financial instability of SCE, some investors have raised concerns that the power suppliers serving the two utilities will not get paid fully, which would impact the stock prices of the power suppliers. Beyond that, the route between supplier and utility can resemble a bowl of spaghetti, because trading contracts often include power that has been sold up to ten times before it reaches a buyer.

In their own defense, generally speaking, power suppliers claim that their increased earnings are due to the fact that corporate operations, including the purchase of power plants and sales of electricity across the country, have grown significantly over the last year. In other words, they are producing and selling more power than the year before, and not just in California. For instance, in 2000, Duke Energy's California power plants generated 70% more electricity than they did in 1999. During the year, the plants provided 17 million MWh of electricity to the state's power grid versus 9.5 million MWh the previous year.

In addition, some of the power suppliers have reiterated that they are working toward creating increased power supply in California, which would help to bring prices down. Again, Duke Energy has been one of the most vocal power suppliers that has gone on the defensive regarding claims of price gouging, and has repeatedly reiterated that it is working to build new plants in California. "We are being vilified, and yet we're the ones being looked to for solutions," said a Duke spokesperson. Duke has proposed to Gov. Davis that it be permitted to use its available resources to build two new power plants in the state, one 500-MW plant that would be ready for commercial operation in 2001 and another 1,500-MW plant that would be ready for 2002.

Even so, presently about a half-dozen lawsuits are still moving forward that accuse electricity wholesalers of unlawfully manipulating the market, and even more suits are expected. Some of the power suppliers have responded that the lawsuits are "a waste of time" and fully expect to be vindicated. Yet some of the plaintiffs, including the city and county of San Francisco, have asked courts to access the generators' internal records, including e-mail and telephone logs. They hope to find a "smoking gun" that will prove market manipulation had occurred in California.

In Conclusion

It is important to note that even one of these factors occurring alone might have wreaked havoc on California's electricity market. When all eight conditions collided and intermingled over the course of nearly two years, the end result was a disastrous circumstance that severely wounded California's energy market and most of its participants.

Chapter 3
Impacts: How the Crisis Manifested

The California energy crisis was an insidious phenomenon that began slowly, but eventually impacted most of the key organizations and companies involved in the state's energy market. The first indications that the state's deregulation electric market was dysfunctional appeared in early 2000, and many reverberations remain unresolved as of December 2001. Certainly, media within the state of California closely tracked the development of the crisis from its onset. National media—even within the energy industry itself—acknowledged the story later, when the full economic impact and threat of power outages had become undeniable. In this chapter, we will address the various tangible impacts that illustrate how the California energy crisis manifested.

SAN DIEGO GAS & ELECTRIC CUSTOMERS BECAME THE FIRST CASUALTY

One of the first indications that California's electric restructuring program was not working properly occurred in the service territory of SDG&E, the smallest of the three California IOUs. In June 1999, the CPUC approved SDG&E's proposal to end its rate freeze on July 1, 1999. The end of the transition period for SDG&E came two and half years early, as SDG&E sold its power plants substantially above book value and thus completed the recovery of its stranded costs. However, unfortunately for SDG&E customers, the end of the rate freeze exposed the utility to sky-high prices for power on the wholesale market, which SDG&E was forced to purchase through the California PX. With the rate freeze lifted, SDG&E began charging its customers market-based rates, which in most cases were double and triple what they were under the rate freeze. While SDG&E paid off its stranded costs and lifted the rate freeze for its customers, the rate freeze remained intact in the service territories of PG&E Co. and SCE.

The jolt of the sharp increase in electricity bills actually came after SDG&E customers received a rebate from the state after the utility paid off its stranded investments. In August 2000, SDG&E customers began receiving deregulation-related checks totaling $390 million. The typical residential customer received $260 and the typical small-business customer received $870. During August 2000, SDG&E sent checks to its 1.1 million residential and small business customers throughout San Diego and southern Orange counties.

The explanation for the rebate is as follows. Under California's restructuring law (AB 1890), the three IOUs in the state issued bonds—called "rate-reduction bonds"—in December 1997 to refinance their debt related to "stranded" assets, which typically included capital that had been previously invested in the construction of generation plants. The issuing of such bonds was known as "securitization", because the issuing of the bonds was secured by a guaranteed cash stream for the IOUs, based on a surcharge

known as a competitive transition charge (CTC) to be paid by electric customers). SDG&E customers had been paying this CTC since competition
began in California in April 1998. The IOUs received approval to issue these
bonds based on their agreement to reduce rates by 10% for their residential
and small-commercial customers.

SDG&E was able to recover its stranded costs in just two and a half
years, well in advance of the scheduled end of the recovery period (March
31, 2002). SDG&E's cost recovery was due in large part to the utility's sale
of its two fossil-fuel power plants at greater than book value. Because SCE
and PG&E Co. were still recovering their stranded costs, which were more
than double the amount of SDG&E's, they were not part of this dividend
payout. Only SDG&E customers as of July 31, 2000, were eligible to receive
this dividend.

Because SDG&E paid off its stranded costs earlier than expected,
customers of the utility should have continued to see savings of approximately
5% off their "base" electricity rates—the portion of their bills that pays for
regulated delivery service. That all sounded like great news for SDG&E
customers, but in reality things weren't very rosy. First, buried in the
announcement of this dividend was the disclosure that SDG&E customers
would also continue to carry a line-item to pay for the bonds over their eight-
year lifetime because they were noncallable. Thus, although SDG&E customers were getting this one-time payment—which again amounted to
approximately $260 for the average residential customer—they would
continue to pay for the bonds for another six years or so.

Second, and more important, was the fact that this dividend paled in
comparison to the substantial increase in the cost of electricity that SDG&E
customers suddenly discovered in the summer of 2000. Now that SDG&E
had paid off its stranded assets, the San Diego area was supposed to resemble
a competitive marketplace, which was intended to provide energy customers
in the area with savings of approximately 20%. This did not happen. In fact,
the "free market" that occurred in San Diego resulted in customers paying
more than they had over the previous two years due to an acute hot weather
peak demand, exacerbated by a tight generation supply within the state.

As a result of these dynamics, the retail market in San Diego, driven in large part by wholesale market conditions, became extremely tumultuous during the summer of 2000. The average electricity bill in the San Diego area doubled during the course of summer 2000, providing the first indication of the effects of free-market forces on the state's energy sector. Further, electricity costs increased between 70–200% for SDG&E customers during that three-month period (before the state was forced to step in and institute an additional rate freeze). In real terms, this meant that some businesses saw their electricity bills increase from $3,800 a month to $10,000 a month, while some residential customers saw their bill increase from around $55 a month to about $103 in the summer of 2000, with further increases expected into the early autumn of that year. There were very few exceptions among energy customers in the San Diego area during that timeframe; residential customers, small businesses, schools, and hospitals all saw their electric bills double.

Gov. Gray Davis entered the fray and called upon federal and state regulators to lower price caps on the state's electricity rates and refund some $100 million to San Diego ratepayers. In July 2000, San Diego's Board of Supervisors voted unanimously to declare a state of emergency because of expensive electricity bills, and heavily petitioned state regulators to reinstate the rate freeze that had been in place.

Within the broader context of the California energy crisis, San Diego represented the extreme problems inherent within the California system, and launched the intense criticisms against legislators, regulators, and the IOUs for not fulfilling their promises of substantial savings from deregulation. The fact that SDG&E paid off its stranded costs earlier than expected, and that the utility's customers could enter the competitive electric power arena sooner than later, would have been a positive if real savings were possible for these energy customers. However, serious flaws in California's deregulation plan were not resolved, which in essence made SDG&E customers the guinea pigs for what became revealed as a failed experiment.

In September 2000, the California Legislature approved a pair of measures to use $150 million in taxpayer money to help San Diegans with that summer's electricity price spikes and to immediately cut electricity bills in San Diego. Basically, this is what the bill proposed: $150 million, collected from California taxpayers, would be set aside until 2003. If at that time, it is determined that SDG&E has encountered financial losses so severe that to cover the losses would mean an additional 10% increase on customer bills, this reserve fund would be used to compensate SDG&E customers. In addition, the bill called for an immediate cut in existing SDG&E electricity rates for all but the largest customers of the utility. Reports indicated that the average residential bill in San Diego would drop from $128 to $68 as a result of this action.

Within this emergency measure was a clause that put a rate cap of 6.5 cents/kWh on SDG&E rates for residential, small commercial, and lighting customers. The bill mandated the CPUC to initiate a voluntary program for large commercial, agricultural and industrial customers of SDG&E to also set the energy component of their bills at 6.5 cents/kWh with a true-up after one year. Therein existed the justification for having taxpayers create the $150 million fund.

On September 8, 2000, Gov. Davis approved emergency legislation to cut electricity rates in San Diego, but did not sign a measure that would have established the $150 million fund, to be paid by taxpayers, to compensate energy customers. Davis did sign the component of the bill that mandated the CPUC to initiate a voluntary program for large commercial, agricultural, and industrial customers of SDG&E to also set the energy component of their bills at 6.5 cents/kWh with a true-up after one year. The rate freeze remains in effect until June 1, 2002, although the option exists to extend it through December 2003 if the CPUC finds that it is in the public interest.

SDG&E, which had urged Davis to veto the legislation, responded by saying that the law was a "quick political fix" that would do nothing to repair severe structure problems in California's wholesale power market. SDG&E estimated that the under-collected shortfall it would bear because of the rate cap would amount to $664 million by the end of 2002.

POWER OUTAGES

The second major indication that California's electric power restructuring program was not working was the increase of power emergencies issued by the California ISO, indicating that power supplies had fallen to dangerously low levels. Since the end of 1999, California experienced an unprecedented number of emergency conditions that in some instances necessitated rotating blackouts. For reference in the subsequent discussion, it is important to understand the definitions of the emergency conditions identified by the California ISO.

Stage One Emergency: Issued when operating reserves reach or begin to fall below 7%. A Stage One Emergency results in public alerts and appeals to conserve power usage.

Stage Two Emergency: Issued when operating reserves are forecast to be less than 5% after dispatching all resources available. A Stage Two Emergency may result in cuts of power to interruptible users. A Stage Two Emergency is only one step short of mandatory rolling blackouts for all customers, a move that would interrupt everything from home air conditioners to traffic lights in the nation's most populous state.

Stage Three Emergency: Issued when operating reserves fall below 1.5% after dispatching all resources available. Involuntary curtailments of service to customers, including rotating blackouts, may be possible during a Stage Three Emergency. A Stage Three Emergency also gives the state of California access to extra megawatts of power from the federal power system.

For comparison, the California ISO issued only one Stage Three emergency in 2000, but had issued a total of 38 Stage Three emergencies as of May 22, 2001. By the same token, Stage One and Two notifications increased from 91 in 2000 to 127 through May 22, 2001.

Data from mid-August 2000 can be used an example of the intense pressure that remained on California's grid system for most of that summer. For instance, on August 16, 2000, approximately 43,218 MW of power was consumed, topping the peak of 43,087 MW that had been reached the day before. Exacerbating the already tight supplies was the fact that some 2,220

MW of generation, mostly in Southern California, was unavailable due to power plant mechanical failures.

As a result, the California ISO issued a Stage Three emergency alert, the highest level of emergency, indicating that power reserves had fallen below 1.5%. In addition, PG&E Co. implemented blackouts across Northern California, including San Francisco and the Silicon Valley. SCE turned off the lights in neighborhoods around Los Angeles. Together, the two utilities serve about 24 million Californians, and operated the rolling blackouts by moving through a series of predetermined "blocks" of customers. Operators at substations were directed to turn off power for a block of customers for about 60 to 90 minutes, then restore service and move on to a subsequent block of customers. Public safety outlets such as police and fire stations were not included in the rolling blackouts, although other public services such as traffic lights were affected by the outages.

Also consider the following examples:

- Between May and December 2000, California experienced 31 days in which a Stage Two alert was issued, indicating that the state's reserves had fallen below 5%.

- On August 8, 2000, the California ISO declared its 14th Stage Two electrical emergency of the year—the fourth consecutive alert in one week—as its operating reserves again dipped below 5%.

- On January 18, 2001, nearly two million California homes and businesses were affected by a second day of blackouts. The blackouts stretched from the Bakersfield area of central California to Oregon, 500 miles away, and lasted for about two hours. Areas impacted included the Bay Area, Napa Valley, Silicon Valley, Sacramento, Modesto, and Turlock. Reportedly, at that time the state came within 1,300 MW of ordering the first statewide blackouts since World War II.

- On January 22, 2001, limited areas of Northern California were blacked out after demand for electricity overwhelmed power grid operators for the third day in less than a week. The outages affected up to 75,000 customers in Sacramento, Roseville, Turlock, and

Modesto areas. At that time, the California ISO directed the California IOUs to institute voluntary load curtailment programs for certain customers within their service areas.

- February 14, 2001, marked 30 days of Stage Three Emergencies. At that time, the California ISO reported that an additional 400 MW of generation became unavailable in Southern California due to off-line generation.

- February 23, 2001, marked the first time since January 13, 2001, that California had not been under a power emergency. Prior to that, the state experienced 32 straight days of Stage Three emergencies.

- On February 28, 2001, the California ISO issued a Stage Two Emergency. At that time, several significant generation sources—including one in Northern California, one in Oregon, and one in Montana—were lost, which prompted the emergency alert. A total of 10,600 MW of California power plants were unavailable due to planned and forced outages and a limited amount of imports from the Northwest.

As will be discussed in chapter 10, power outage warnings continued into the spring of 2001, predicting that the summer of 2001 would be fraught with power outages in the range of 55 hours to a high of 700 hours. Due to a fortunate combination of factors such as increased conservation efforts, mild weather patters and new generation facilities that quickly came online, the reliability of California's power market improved significantly when comparing summer 2001 to summer 2000.

THE RISE IN WHOLESALE ELECTRICITY PRICES

The third indication that energy deregulation in California was resulting in a full-blown crisis situation was the sharp rise in wholesale power prices. Wholesale prices for electricity in California, sold through the California

PX, embarked on a dramatic increase in the early summer of 2000, due to factors that included a growing shortage of natural gas and increased demand in the state. The trend would continue throughout the remainder of 2000. For instance, from June 2000 through July 2000, wholesale electricity prices increased on average 270% over the same period in 1999, according to the CPUC. By December 2000, wholesale prices on the California PX cleared at $376.99/MWh, over 11 times higher than the average clearing price of $29.71/MWh in December 1999, according to records from the California PX.

In fact, according to a December 5, 2000, report in *The Wall Street Journal*, natural gas prices, which already had been rising all year, shot up to all-time highs on the New York Mercantile Exchange (NYMEX). In the early part of December 2000, natural gas future prices were hovering around the $8.6/MMBtu for January 2001 delivery at the Henry Hub trading exchange. The increase resulted from a volatile mixture of rising demand and decreased supplies for natural gas, which had driven up prices by a total of 218.9%, or more than $5/MMBtu, since December 1999. There were even reports circulating that spot (cash) prices at the Southern California border reached an astounding $41/MMBtu on December 4, 2000, breaking the Daily Gas Price Index's all-time record high of $39/MMBtu. The sharp increases were viewed as a reaction to cold weather in the Northeast and forecasts of more to come. At the time, projections continued to indicate that natural gas customers would see sharp increases in their heating bills over the winter months.

According to information from the California PX, generators and marketers were charging as much as $853/MWh in June 2000, an astounding increase over prices as low as $1/MWh the previous year.

Wholesale prices for electricity in California exceeded $1,200/MWh in June 2000, compared to only $50/MWh at the same time in June 1999.

The Wall Street Journal article on the price spikes also reported that the American Gas Association estimated that, as of November 24, 2000, there were about 2,502 trillion cubic feet (Tcf) of gas in storage. This represented an 11% decrease from the five-year average for that date. The United States is mostly self-reliant when it comes to natural gas supplies, and the fact that a reported 60% of federal lands had been placed off limits for natural gas exploratory drilling certainly added to the concern about future supplies. Environmental groups and the administration under President Bill Clinton were fiercely protective of Federal Reserve lands and prohibited drilling for natural gas (or oil) in states like Alaska, where some 36 Tcf of gas are believed to be locked underground.

Yet, at the same time, substitute forms of power generation, such as coal and nuclear, had been restricted or downgraded in favor of natural gas, which had only exacerbated the strained demand on supply. Environmental Protection Administration standards had become very pro-natural gas over the previous decade, and as a result few coal-fired plants had been built in the last 10–20 years. Certainly, the public resistance toward nuclear power has been well documented, which also thwarted any planned production of a new nuclear plant.

Wholesale spot prices for electricity skyrocketed in the days immediately after Dec. 8, 2000, the day when FERC ordered the California ISO to lift its $250/MWh cap. FERC instead imposed a "soft" price cap under which any deals exceeding $250/MWh would be subject to a refund if they were found to be excessive. On December 12, the peak spot price for power through the PX shot up to $988/MWh. This compared to an average electricity price of $30/MWh a year earlier. Also on December 12, natural gas, which had sold at $3/MMBtu a year prior was selling at $60/MMBtu, a 2,000% increase.

The average gas price in California more than quadrupled during the fourth quarter in 2000 compared to the fourth quarter 1999, to $12.63/MMBtu. Prices rose as high as $53.38/MMBtu in early December 2000, according to Bloomberg Energy Service statistics.

In June 2001, much lower demand and a flood of supply on the California gas pipeline grid led to sharply lower spot natural gas prices, especially at the California border, where prices dropped $5-$6/MMBtu. Mild temperatures and more power coming into the state from the Pacific Northwest took most of the credit for the price decreases. Wholesale prices at the California border plummeted to lows around $3.20/MMBtu at the Topock, AZ, connection into Southern California Gas. The previous Friday (June 1, 2001), the Southern California Border average was $9.24/MMBtu.

The sudden drop in natural gas prices marked the first time in nearly a year that spot market averages had remained in the $4/MMBtu range. In fact, prices in the range of $3.50 to $4/MMBtu had not been tracked since July 2000. While the drop was anticipated as a result of stabilization in the market, an interesting dynamic to the falling prices was the extent to which the industry paradigm had shifted with regard to the industry's definition of "low prices." Yes, the $4/MMBtu mark seemed low compared to the 12-month average, but it was still high when compared to the previous 10 years.

The big question to be addressed is how long the "low" gas prices will continue. The jury is still out on this debate, and there is little consensus on this question. Further, the issue of natural gas prices remains very subjective. Naturally, those in the natural gas-production business would like to see prices rise once again, while those on the purchasing side believe the current fall in prices below the $4/MMBtu range is not low enough. Along with this value discrepancy, debate also continues over how long gas prices will remain below $4/MMBtu, with many analysts arguing that the drop is only temporary.

Another result of the drop in natural gas prices is the increase in storage supplies. Traders expect to see a large build reported in the American Gas Association's storage reports to be released later this month. Early predictions suggest a national build of 90 Bcf to as high as 120 Bcf, which is considered bearish for the gas market.

On the other hand, market forecasters have identified several factors that could cause natural gas prices to rise again. For instance, prices could jump significantly during subsequent summer seasons as insufficient pipelines become clogged once again with rising demand (especially in California).

Further, the dramatic increase in new natural gas-fired power plant production that is anticipated across the country (a plan that has been endorsed by the Bush administration) will cause demand to increase substantially, driving up gas prices once again. Most of the 15,000 MW worth of new power plants that are in the planning stages will be fueled by natural gas, and consequently should significantly increase energy demand across the country. Demand and high prices for natural gas naturally go hand in hand. With few potential alternatives, natural gas is the fuel of choice for new power plant construction. This can only put upward pressure on gas prices as the demand increases.

In addition, the ongoing drought in the Pacific Northwest will continue to compromise hydroelectric supplies in that region, which could further exacerbate the region's demand for gas-fired generation. Analysts suggesting that California should not get used to the lower gas prices further point to the fact that the state still has an 18% year-on-year deficit in regional inventories, which keeps the region's natural gas supply in a compromised position.

FINANCIAL INSTABILITY OF PACIFIC GAS & ELECTRIC AND SOUTHERN CALIFORNIA EDISON

Chapter 5 of this book is devoted entirely to PG&E Co.'s path to bankruptcy proceedings and the ongoing financial vulnerability of SCE. However, it is important to note that for many observers of the California energy crisis, the financial problems of the two largest IOUs in California—and most certainly the well-publicized bankruptcy of PG&E Co.—may have been the first indication of the severity of the problems with the California energy market.

On April 6, 2001, PG&E Co. filed for bankruptcy protection under Chapter 11 of the U.S. Bankruptcy Code. At that time, its parent company PG&E Corp. estimated that the utility had spent some $9 billion for wholesale power with no opportunity for reimbursement for those expenditures due to the rate freeze that was still in effect for the utility.

Although most of the publicity related to the price spikes in the summer of 2000 focused on SDG&E, the legacy of the market problems in California was arguably carried by PG&E and SCE. Both utilities had not even fully entered competitive markets within the state, and yet bore financial ramifications from the dysfunctional market.

As noted earlier in this text, PG&E Co. was forced to purchase its power on the wholesale market. In addition, under California's restructuring mandates, PG&E Co. had to buy its power out of the PX rather than establishing bilateral contracts with suppliers. The combination of low supply and high demand drove up the wholesale cost of electric power, which PG&E Co. faced as a buyer. Due to the retail rate freeze remaining in effect, PG&E Co. was unable to pass on the high cost of electric power to its customers. Although reports varied, on average PG&E Co. had been paying up to 19 cents/kWh on the wholesale market, when by law it is only allowed to charge its customers on average five cents/kWh.

PG&E Co. said it was taking the bankruptcy action in light of its uncollected energy costs, which had increased by more than $300 million per month; continuing CPUC decisions that economically disadvantage the company; and the unmistakable fact that negotiations with Gov. Gray Davis and his representatives are going nowhere. Neither PG&E Corp. nor any of its other subsidiaries, including its National Energy Group, have filed for Chapter 11 reorganization or are affected by the utility's filing.

The problems of SCE, while less extreme and less publicized than those of PG&E Co., nevertheless kept the utility in financial uncertainty for most of 2000 and 2001. While PG&E Co. took the approach of declaring bankruptcy, SCE opted to continue negotiating a rescue plan with California officials, which it more or less achieved toward the end of 2001.

HUGE PROFITS FOR GENERATORS

While the ethical debate over whether power supplies have engaged in any unscrupulous (or illegal) business practices rages on, the numbers released by such companies paint a pretty clear picture that California's energy market to one extent or another became a very profitable market for those energy companies that sold power to the state. California in particular and a national imbalance between supply and demand in general resulted in a financial windfall for those companies in the business to generate and/or deliver electricity to power-starved regions across the United States.

The following companies were identified as having—to one extent or another—contributed to what federal regulators deemed to be unjust and unreasonable wholesale prices in California.

- Arizona Public Service
- Avista Energy
- Calpine Corp.
- Duke Energy Trading and Marketing
- Dynegy Power Marketing
- El Paso Energy
- Enron
- Mirant Corp.
- Nevada Power Co.
- Portland General Electric Co.
- Public Service Co. of Colorado
- Public Service Co. of New Mexico
- Reliant Energy
- Williams Energy Services

Certainly it has been Gov. Davis' position that power generators and markets active in the California, particularly a small group of Texas-based companies, charged inflated prices for power sales to the state, which by some estimates resulted in nearly $9 billion in overcharges. Davis based much of his claims regarding over-charges on a California ISO report, first released in March 2001, which stated that power sellers had earned $6.3 billion in excess profits between May 2000 and February 2001. In fact, the report, which later revised its figures to $6.7 billion, became a crucial element of the Davis administration's campaigns against alleged electricity price gougers.

Of the companies singled out, according to data released by the California ISO, Oklahoma-based Williams Companies led the power companies in charges to California with $860 million, followed by Duke Energy with $805 million, Southern Company Energy Marketing (now known as Mirant) with $754 million, and Reliant Energy Services with $750 million.

As a whole, power generators began to show a dramatic increase in earnings starting with year-end 2000, and particularly in the first and second quarters of 2001. Many in the group posted earnings growth of as much as 20% during the first quarter of 2001, which was one of the first financial quarters to reveal the extent to which energy generators and traders had profited from the California market.

Let's examine the financial performance of some of the companies involved in energy marketing in California.

Calpine

Calpine Corp. (NYSE: CPN) reported 2000 net income, before extraordinary charge, at $324.7 million, representing a 238% increase over 1999. For the first half of 2001, Calpine earned $227.3 million, or 68 a cents a share, on revenues of $2.3 billion, compared with net income of $80.6 million, or 28 cents a share, on revenues of $548 million for the first six months of 2000. According to company officials, Calpine's success was attributed to long-term contracts it negotiated during the previous year when wholesale prices remained high. Several of those contracts were in California, where Calpine is "doing very, very well," according to the company CEO. For instance, Calpine charged an average of $97.64/MWh for the 2.5 million MWh that it produced for use in the West during the second quarter of 2001. In contrast, the company charged an average of $63.99/MWh for two million MWh produced in the East and an average of $50.27/MWh for three million MWh in Central U.S. markets.

Duke

Duke Energy reported a net income of $1.25 billion, or $3.40 a share, for year-end 2000. For the second quarter of 2001, Duke posted a 27% increase in its net income, with net rising to $419 million, or 53 cents a share, from $329 million, or 44 cents a share in the second quarter of 2000. Revenue in the second quarter of 2001 rose 43% to $15.58 billion from $10.93 billion. In total, Duke's businesses outside of its traditional regulated business in North Carolina, including its operations in California, accounted for more than 60%

of $917 million in operating earnings in the second quarter of 2001. For the third quarter of 2001, Duke said that its third-quarter profits rose 46% due to the strong performance of its merchant generation and trading and marketing units. Looking ahead, the company said it expects to report 2002 earnings near the high end of its guidance of 10–15% earnings-per-share growth from its 2000 earnings of $2.10 a share. The current consensus estimate is $2.81, according to research firm Thomson Financial/First Call.

Dynegy

Considered one of the major power generators in California, Dynegy said it reported a 210% increase in 2000 recurring net income to $452 million, or $1.43 per diluted share, compared to 1999 recurring net income of $146 million, or $0.63 per diluted share. The 2000 results represented a 157% increase compared to 1999 *pro forma* recurring net income of $176 million, or $0.53 per diluted share. For the year, Dynegy's common stock was one of the top performers among Standard & Poor's 500 companies, with a total shareholder return of 218%. In addition, Dynegy reported a 48% increase in second quarter 2001 recurring earnings per diluted share to $0.43, compared to second quarter 2000 recurring earnings per diluted share of $0.29. Recurring net income increased 60% in second quarter 2001 to $146 million, compared to second quarter 2000 recurring net income of $91 million. In the third quarter of 2001, Dynegy said that it gained a 55% increase in its recurring earnings per diluted share to $0.85, compared to third quarter 2000 recurring earnings per diluted share of $0.55. Recurring net income increased 62% in third quarter 2001 to $286 million, compared to third quarter 2000 recurring net income of $177 million.

Enron

Houston-based Enron Corp. announced record financial and operating results for the full year 2000, including a 25% increase in earnings per diluted share to $1.47; a 32% increase in net income to $1.3 billion; a 59% increase in marketed energy volumes to 52 trillion Btu equivalents per day; and an almost doubling of new retail energy services contracts to $16.1 billion. Enron also announced a very successful fourth quarter of 2000, generating recurring earnings of $0.41 per diluted share, an increase of 32% from $0.31 in 2000. For the third quarter of 2001, Enron Corp. reported recurring earnings per diluted share of $0.43 for the third quarter of 2001, compared to $0.34 a year ago. Total recurring net income increased to $393 million, versus $292 million a year ago. Enron also reaffirmed it is on track to continue strong earnings growth and achieve its previously stated targets of recurring earnings per diluted share of $0.45 for the fourth quarter 2001, $1.80 for 2001, and $2.15 for 2002.

Note that Enron declared bankruptcy on December 2, 2001, following a terminated agreement from Dynegy Corp. to acquire the company.

Mirant Corporation

Mirant (NYSE:MIR), a former subsidiary of Southern Company known as Southern Energy until January 2001, reported net income for year-end 2000 of $332 million. Mirant's earnings from operations for year-end 2000 were $336 million, which represented a 36% increase from 1999 earnings from operations of $270 million. As one of the top natural gas and power-marketing firms in the nation, Mirant moved more than 6.9 Bcf of natural gas per day and sold $88 million MWh of power in 2000. The same success has continued into 2001 for the company. In the third-quarter of 2001, Mirant reported record earnings of $234 million, or 67 cents per diluted share, compared with $119 million, or 35 cents per diluted share, in the third quarter of 2000. While Mirant Corp. has not singled out its California profits in any of its financial reports (much like many of the other companies listed), it is one of the companies that has formed a long-term power contract with the California Department of Water Resources and was active in the state's spot market activity.

Reliant Energy

Reliant Energy (NYSE: REI) reported third quarter 2001 adjusted earnings of $312 million, or $1.07 per diluted share, compared to adjusted earnings of $396 million, or $1.37 per diluted share, for the third quarter of 2000. For the nine months, which ended September 30, 2001, Reliant Energy's adjusted earnings rose to $833 million, or $2.86 per diluted share, compared to adjusted earnings of $765 million, or $2.67 per diluted share, for the same period in 2000. Reliant officials acknowledged that the increase in adjusted earnings for the year to date was largely driven by improved performance from the company's wholesale energy segment, partially offset by a decline in operating income from the electric operations segment due to substantially milder weather this year and a decline in the European energy segment.

Specifically, the company's wholesale energy segment produced third-quarter operating income of $266 million, compared to $314 million for the third quarter of 2000. The decline was primarily the result of lower gas and power unit margins in 2001 compared to the third quarter of 2000, when unique market conditions and favorable hedging produced exceptional results. Also contributing to the decline were increased plant operation and maintenance expenses related to Western operations, higher legal and regulatory expenses related to Western markets, and higher administrative costs to support the company's growing wholesale commercial activities and operations.

Williams

In 2000, Williams Companies earned more than $1 billion from its energy trading operations alone—an 870% increase over the previous year's numbers. Ironically, Williams does not own any of the power plants that had been sold by California's IOUs, but the company is active in the state's wholesale trading market. Actually, Williams had formed a contract with AES Corp., which had purchased some of the California plants. By 1998, AES had signed a contract that would result in more than $1 billion worth of power profits for Williams. Under the 20-year agreement, Williams essentially pays

AES to convert natural gas into electricity, which Williams then markets to buyers, including those in California. Williams became singled out among the many power generating companies that sold power in the California market. In fact, Gov. Davis on more than one occasion claimed that Williams alone owed the state $861 million—more than any other out-of-state energy company. For year-end 2000, Williams reported unaudited 2000 income from continuing operations of $873.2 million, or $1.95 per share on a diluted basis, versus $178 million, or 40 cents per share on a restated basis, for 1999.

According to the company, the increase primarily was due to substantially higher profits from the energy marketing and trading business, reflecting successful proprietary natural gas and electric power trading during a year of nationwide volatility across these energy portfolios, increased earnings from structured transactions, and greater overall market demand. Also contributing were increased natural gas liquids margins and volumes, higher natural gas production prices, and increased throughput and refining margins.

For the third quarter of 2001, Williams reported unaudited third-quarter net income of 44 cents per share on a diluted basis vs. 27 cents per share for the same period of last year. Absent investment write-downs and other charges, Williams' recurring third quarter 2001 results would be 65 cents per share, compared with the Wall Street consensus estimate of 53 cents.

Such power suppliers have been intensely criticized by some in the industry, including California state officials, who have accused them of price gouging and unjustly profiting from the California crisis.

In March 2001, the California ISO concluded that electricity wholesalers had charged excessively high prices for power in the first few months of 2001—roughly $550 million—and should be forced to refund the money. The refund order would have to be issued by FERC, but the California ISO was establishing its perspective that such refunds would be warranted. To determine if wholesalers were gouging grid managers buying power on the spot market, the California ISO estimated what it determined to be reasonable costs for each plant, and then asked each plant's owners to provide further information justifying the wholesale prices that they charged. Many of the generators refused to supply this information, stating that it was competitive and proprietary data. Duke Energy was one company that immediately responded to the California ISO report, arguing that it had not overcharged for power. Duke claimed that it sells most of its power into California through long-term contracts, and thus only a small percentage of its power was sold into the California spot market.

Interestingly, according to a report in the *Los Angeles Times* in July 2001, during the first three months of 2001, an assortment of public and private entities charged California prices averaging well above some of those paid to Texas firms, according to documents that the newspaper had obtained from the Department of Water Resources. According to the report, while the average prices charged by Dynegy, Duke, and Mirant ranged from $146/MWh to $240/MWh, prices charged by the Canadian public utility BC Hydro, the LADWP, and Sempra Energy, an affiliate of SDG&E, ranged from $292 /MWh to $498/MWh. According to the same report, in the first quarter of 2001, some public entities such as LADWP and BC Hydro far exceeded those of the biggest private companies. For example, Enron charged an average of $181/MWh and Mirant charged an average of $225/MWh. Meanwhile, Calgary, Canada-based TransAlta Energy, which also sold into the California market, charged an average of $335/MWh.

SUCCESS OF LADWP

While most of California suffered from repeated power emergency warnings and the state's IOUs struggled financially, the municipal utility known as the LADWP emerged as a real success story. Amid consistent power outage warnings and utilities strapped with the financial debts, the municipal utility that primarily serves the greater Los Angeles metro area announced that its more than 3.7 million businesses and residences were enjoying stable electricity prices and a plentiful supply.

What a sweet irony this must have been for S. David Freeman, LADWP's former general manager. (Note that Freeman left LADWP in 2001 after being appointed chairman of the new California Consumer

Power and Conservation Financing Authority). In 1998, LADWP—the largest city-owned utility in the United States—was ridden with massive debts and faced an uncertain future. In many ways LADWP now can be seen as the only success story on a landscape that is marked by one IOU maelstrom after another.

Let's take a step back and assess how LADWP was able to insulate itself from the California crisis. In 1997, Freeman was brought on board at LADWP (from the Cal-ISO, no less) to essentially save what was thought to be a sinking ship. Straddled with $7.9 billion in debt, the need to cut about 1,500 jobs, and a likelihood of increasing rates for both residential and industrial customers, Freeman's decision to take on the role of managing LADWP seemed like many to be a herculean task. Yet Freeman arrived at LADWP with an aggressive reorganization in mind, including the primary goal of lowering rates and becoming debt-free by 2003.

Early on, Freeman made a critical decision—one that in hindsight appears to be flawless—for LADWP to opt out of deregulation while the state's three IOUs continued down the treacherous path of restructuring. While the IOUs began divesting themselves of generating facilities as means of averting market power issues (due to agreements established with the CPUC), LADWP was a prime example of public power throughout the United States electing to keep the integrated utility model. In addition, Freeman blocked ongoing attempts by a consortium led by Duke Energy to buy LADWP's trading unit, and thus maintained control over the municipal's generation assets.

Enter the summer of California's discontent. With demand on an unexpected increase and supplies in questionable availability, LADWP found itself having the upper hand. The wholesale market clearly had favored generators, and LADWP was one of an elite group of companies that capitalized on the disparity between supply and demand. Critics accused LADWP of "profiteering," but Freeman contended that the utility was just taking advantage of California's flawed market and was able to sell its excess generation back to the California market. The truth is that LADWP was in the right place at the right time, and maximized its ability to generate cash via power trading, which it used to drive down its debt. This was a shrewd move by any estimation.

The municipal utility ran its plants flat out for most of the hot season of 2000, illustrated by the fact that it got fined for violating emission standards. Some reports indicated that LADWP earned about $45 million just from surplus power over the course of the summer. These profits only added to what LADWP had made in the previous year. Fiscal year 1999 wholesale energy marketing and transmission services generated $98 million, which, along with other sources of wholesale revenue, marked a growth of 118% over 1998.

And, while the state government evaluated long-term contracts with energy providers (based on high prices), LADWP was free to continue with an aggressive generation plan. Among its objectives are to add 2,900 MW of power generation within its service territory in the Los Angeles basin. The end result of this would be less dependence on external sources for power and instant supply available at times of peak demand.

Along with having no worries about supply, LADWP customers also can look forward to a 5% rate reduction that should take effect in 2002, followed by a 10% rate reduction that has been proposed for 2003. As criticism mounted against California's competitive wholesale market, it certainly appears that LADWP took a prudent route by opting out of competition and retaining its integrated model three years ago.

Chapter 4
The Financial Instability of Pacific Gas and Electric Co. and Southern California Edison

While electricity generators began charging record power prices in 2000, PG&E Co. and SCE began accumulating twin piles of debt because AB 1890 had frozen the retail rates charged to households and small businesses. In addition, the utilities were forced, also through AB 1890, to purchase the vast majority of their power supply needs through the California PX, a spot market that reflected the skyrocketing cost for power throughout 2000. The combination of high wholesale costs and the inability to pass the costs on to customers due to a rate freeze led both PG&E Co. and SCE into dire financial straits, a condition that neither utility has fully resolved as of this writing. The third IOU in California, SDG&E, averted financial problems mostly because it had paid off its stranded costs earlier than its two counterparts, and thus was able to pass along the high costs of wholesale power to its customers.

PG&E Co. and SCE, which serve 24 million Californians, both petitioned state regulators for the right to raise rates in 2000, but were rebuffed. Consequently, both utilities ran out of cash by mid-January 2001 and could not find a generator willing to continue to sell them electricity, forcing the state to step in and begin buying power for Californians on behalf of the IOUs. Interestingly, the two utilities took very different paths toward attempting to resolve their financial problems. On one hand, PG&E Co. abandoned any chance of working out a resolution plan with the State of California, and declared bankruptcy soon after it became clear that its financial problems had become insurmountable. On the other hand, SCE has steadfastly refused to declare bankruptcy, choosing instead to continue negotiations with the State of California in the hope of reaching some sort of "rescue plan." As of this writing, PG&E Co. continues to navigate through bankruptcy proceedings while its parent PG&E Corp. attempts to protect its other subsidiaries, and SCE continues to remain at the mercy of the California Legislature for an acceptable plan that will keep it financially solvent.

Pacific Gas & Electric Co.

On April 6, 2001, PG&E Co. filed for reorganization under Chapter 11 Bankruptcy Code in the U.S. Bankruptcy Court for the Northern District of California in San Francisco. The company said it was taking this action in light of its uncollected energy costs, which were increasing by more than $300 million per month; continuing CPUC decisions that economically disadvantaged the company; and the unmistakable fact that negotiations with Gov. Gray Davis and his representatives were going nowhere. At the time of the filing, PG&E Co. reportedly had more than $9 billion in uncollected costs for wholesale power purchases. Neither PG&E Corp. nor any of its other subsidiaries, including its National Energy Group, filed for Chapter 11 reorganization or were affected by the utility's filing.

Despite the fact that PG&E Co.'s bankruptcy filing had been in the making for several months, and that its financial insolvency appeared inevitable, the actual filing for Chapter 11 protection caught many by surprise, including California's top officials. Ironically, the utility's 13 million customers may end up being the least impacted by the bankruptcy, as PG&E Co. will remain in the business of providing power while it works its way through bankruptcy court, probably for years. However, there is little doubt that the bankruptcy filing of California's largest utility continues to spark many reverberations throughout the state and across the country.

At the time of the bankruptcy filing, there appeared to be some fairly obvious observations about and ramifications to PG&E Co.'s decision to declare bankruptcy instead of continuing to work with the State of California on a resolution.

Consider the following points.

The bankruptcy was announced the day after it became clear to PG&E Co. that the state of California would not be purchasing its transmission assets at a price that its parent PG&E Corp. believed the assets to be worth.

In addition, a rate increase approved by the governor could not be used to repay any of PG&E Co.'s existing debts. The night before the bankruptcy filing, Gov. Gray Davis delivered a televised address to the state in which he indicated that the rate increases would be contingent on PG&E Co. selling the transmission assets to the state. In other words, if PG&E Co. agreed to the state's price for the transmission assets, a portion of the rate increase would be used to alleviate the utility's subsequent debts. The inside word was that PG&E Co. was finally driven to its bankruptcy decision due to disagreements with Gov. Gray Davis regarding the price of the transmission assets, which could be anywhere from $3 billion to $9 billion. Officially, PG&E Co. said that its negotiations with the state "were going nowhere." The CPUC had previously approved rate increases of up to 46% for the customers of SCE and PG&E Co., but clearly stipulated that none of this added revenue could be used to pay off existing debt. In his statewide address on the eve of the bankruptcy, Gov. Gray Davis reluctantly announced an additional rate increase for non-residential customers that would average about 26.5%. Davis had fiercely resisted any form of a rate increase to accommodate PG&E Co.'s debt. The official PG&E Corp. response to both increases was that they were insufficient to pay off existing debts of the utility operation.

PG&E Corp. and other non-utility subsidiaries are not impacted by the bankruptcy measure (at least for now).

Due to a shrewd financial restructuring that was approved by FERC a few months before the bankruptcy filing, high-profit subsidiaries such as National Energy Group and PG&E Gas Transmission Northwest Corp. were placed in a special-purpose, bankruptcy-remote entity known as GTN Holdings LLC.

Consequently, these non-regulated subsidiaries (along with the parent corporation) may not be impacted by the utility's bankruptcy filing. The U.S. federal bankruptcy judge ultimately may take a closer look at how profits were divided among the parent operation and other non-regulated subsidiaries, but for the time being only PG&E Co. has declared bankruptcy.

Power suppliers still owed money by PG&E could likely never get paid.

Much of PG&E Co.'s staggering debt is owed to power suppliers that sold power to the utility during 2000. With the onset of lengthy legal proceedings, it becomes more unlikely that these power suppliers will be paid what they are owed any time soon. Consequently, power suppliers owed more than $100 million by PG&E Co. also saw their stocks drop as a reaction to the bankruptcy announcement. According to a report in Reuters, on the day of the bankruptcy filing (April 6, 2001) Duke Energy (NYSE: DUK) was down $2.65 or 6.25% at $30.75; Dynegy Corp. (NYSE: DYN) was off $2.58 or 5.07% at $48.34; Williams Companies (NYSE: WMB) was $2.26 or 5.4% lower at $39.60; and Reliant Energy (NYSE: REI) lost $2.70 or 5.95% to $42.65. In addition, there is the real question of whether or not such power suppliers will be willing to provide power used by a bankrupt company (even if the state of California remains involved as the buying intermediary). For its part, PG&E Corp. Chairman Bob Glynn said that the utility subsidiary intends to eventually "pay all our debt" in full, but he did not report a specific figure. In addition to the power suppliers, banks that have provided loans to PG&E Co. (such as Bank of America, Wells Fargo, and J.P. Morgan Chase) could all be impacted by the bankruptcy. PG&E Co.'s preliminary bankruptcy filing lists its top creditor as Bank of New York, to which it owes more than $2.2 billion. Although some payments may be made after years of litigation, such lenders and bondholders may be forced to write off billions of dollars that were previously advanced to PG&E Co. as losses.

The blame game is in full force.

Gov. Gray Davis and state regulators repeatedly blamed out-of-state generators for price gouging and federal regulators for failing to impose wholesale rate caps on electricity prices. Also, Gov. Gray Davis has issued harsh words against PG&E Corp. During the weekend before the bankruptcy filing, Davis appeared on two nationally televised news programs to berate PG&E Corp. for awarding an estimated $50 million in bonuses and raises to about 6,000 mid-level managers and support staff on the eve of the utility's bankruptcy filing. "Management at PG&E Corp. is just focused upon padding their own pockets, not in discharging their duty to serve their many customers in California," Davis reportedly said. Earlier, Davis had previously issued a statement saying PG&E "management is suffering from two afflictions: denial and greed." In addition, California regulators criticized FERC for not implementing wholesale price caps in the Western region, which they believed would help to stabilize the market in California. U.S. Sen. Dianne Feinstein said, "It is inexcusable that FERC has refused to intervene and provide temporary stability and reliability [through the use of wholesale price caps]." Feinstein vowed to introduce legislation to force federally mandated price caps for the entire Western region. For its part, FERC refused to take responsibility for the California crisis, saying instead that PG&E Co.'s bankruptcy filing was "the unfortunate result of a massive failure by policymakers at all levels." PG&E Co. blamed decisions made by the CPUC and Gov. Gray Davis that it believed "undermined" its ability to return to financial viability and recover its uncollected costs. Specifically, PG&E Co. blamed the California government for not allowing it to recover prior wholesale power costs through recent rate increases, which essentially has kept the utility in a debt situation. PG&E also believed that the governor was not moving quickly enough to reach an agreement regarding the sale of PG&E Co.'s transmission system to the state. Consumer groups continue to lambast the utility and its parent company, which they claim had $30 billion in available capital that could have been used to pull its subsidiary out of debt.

Could the bankruptcy filing of SCE be far behind?

In response to the bankruptcy filing of PG&E Co., SCE, the second-largest utility in California, continued to maintain that it intended to avoid bankruptcy, saying that it believed it could work out a "comprehensive solution to our current crisis." However, because the two utilities have been so closely linked through the state's energy crisis, PG&E Co.'s decision to declare bankruptcy certainly raised the odds that SCE would eventually follow suit. Moody's Investors Service reported that PG&E Co.'s bankruptcy increased the possibility of an SCE bankruptcy filing. Further, Moody's reported that numerous creditor groups, particularly the small generators, have become impatient with the time that it has taken to advance settlement discussions with the state of California, thereby increasing the prospects for an involuntary petition by a group of creditors. On April 6, 2001, the day of PG&E Co.'s bankruptcy filing, shares of SCE tumbled $4.29, or nearly 34%, to finish trading at $8.35.

This is just the start of a long, drawn out legal process.

Electric utilities that previously declared bankruptcy (including El Paso Electric and Public Service Company of New Hampshire) found themselves embroiled in legal proceedings for several years. Thus, although the ultimate outcome for PG&E Co.'s bankruptcy proceedings is not presently known, it is fairly clear that the company will be essentially consumed with litigation and a thorough investigation of its financial records for the foreseeable future. Federal Bankruptcy Judge Dennis Montali of San Francisco has been assigned the PG&E Co. case. Montali will have the ultimate power to approve or reject any plan for PG&E Co.'s reorganization or repayment of its debt.

Several positive elements of the process that might work to PG&E Co.'s advantage is that the utility will now be held accountable to only one individual, Judge Montali. PG&E Co. will no longer have to appease several regulatory entities across state and federal lines. Second, by the time the bankruptcy proceedings are completed, market dynamics hopefully will have improved and California's supply problem should be resolved. Consequently, PG&E Co. could eventually rise again, like the phoenix out of its ashes, and emerge as a much stronger company than it is today.

PG&E had taken the wise move of protecting its high-growth assets months before it actually declared bankruptcy. In January 2001, Standard & Poor's lowered its credit ratings of PG&E Co. and other units under the parent corporation to triple B-minus, the rating level one notch above speculative grade. S&P also lowered its ratings on bonds issued by various issuers for the benefit of PG&E Co. S&P issues credit ratings that reflect the organization's independent opinion of the general creditworthiness of a company holding debts, or the creditworthiness of an obligation with respect to a specific debt security. Essentially, the goal of S&P is to provide investors with an independent and objective assessment of a company's financial condition.

The downgrades of PG&E (along with SCE) reflected what S&P believed to be a failure on the CPUC's part to "meaningfully address a market structure that compels the utility to serve customers at prices that are substantially below the cost of procuring wholesale power." The downgrades also reflected the likelihood that PG&E Co.'s eroded financial performance would impair the financial strength of the subsidiaries. Further, S&P stated that PG&E Co.'s resulting heavy debt burden and dearth of liquidity were inconsistent with a higher grade and might lead to further downgrades. S&P stopped short of assigning additional lower ratings in the hope that legislative sessions taking place in California might result in some long-term decisions that would benefit PG&E Co. S&P stated that any prospects for a favorable resolution to the financial crisis of PG&E Co. now lie with the California Legislature. However, if legislative action is not taken quickly, S&P anticipates further substantial downgrades of the PG&E Corp. family. Currently, the vast majority of PG&E remains on CreditWatch with negative implications, reflecting the potential for the ratings to be further downgraded if PG&E's "financial condition is permitted to continue to deteriorate."

What is important to note is that, with one exception, S&P downgraded the credit ratings of the complete family of PG&E Corp, including PG&E Co., the regulated utility, and PG&E Pacific Venture Capital LLC, which manages a portfolio of capital investments in energy and telecommunications companies. The one (very important) exception to this was PG&E Gas Transmission Northwest Corp. (GTN), which had recently been bifurcated from the PG&E Corp. family and placed into a special-purpose, bankruptcy remote entity known as GTN Holdings LLC. PG&E's National Energy Group (NEG) had formed GTN Holdings and contributed to the entity 100% of the issued and outstanding shares of GTN. At the time of the bankruptcy filing, S&P affirmed its single A-minus rating on GTN and removed it from CreditWatch Negative where it had been placed on December 13, 2000. According to S&P, this structure is designed to ring-fence GTN from the possible bankruptcy or credit downgrade of PG&E Corp.

NEG's subsidiary PG&E Energy & Trading also gained bankruptcy immunity under the protection of GTN Holdings. According to S&P, PG&E attempted to secure bankruptcy protection for NEG itself. This is a very important development as NEG represents one of the largest profit generators under the PG&E Corp. structure. NEG is an unregulated, wholly-owned subsidiary and one of the nation's leading competitive power producers.

PG&E has said the sustained earnings performance has NEG on track to meet its goal of delivering 30% of the corporation's earnings by 2002. Thus, it's no small wonder why PG&E Corp. would want to protect this valuable asset from any negative repercussions should it declare bankruptcy as a parent company.

Despite its ardent efforts to protect its other subsidiaries from any impacting stemming from the bankruptcy of PG&E Co, by October 2001 PG&E Corp. was learning that the CPUC remains among the tallest of obstacles standing between it and a speedy resolution of its utility's bankruptcy woes. Lawyers for PG&E Co. met in court on October 9 with a federal bankruptcy judge, state representatives, and lawyers representing thousands of creditors to craft a timeline of the process by which the utility could emerge from bankruptcy and pay its debts. During the nearly three-hour meeting, critics of PG&E Co.'s bankruptcy plan repeatedly asked Judge Montali to delay allowing the utility to file a disclosure statement—an explanation of how it plans to pay debts and restructure its company—until it is clear whether the utility must wait for state approval before moving forward.

The restructuring plan that became apparent by the end of 2001 is a significant step beyond the mere protection of National Energy Group and other subsidiaries under the parent company PG&E Corp. The plan would essentially create two (possibly three) new companies: one for natural gas transmission, one for electrical transmission, and one that would own PG&E Co.'s hydroelectric and nuclear power plants. All of these companies would come under the corporate umbrella of PG&E Co.'s parent company, PG&E Corp., and would be operated separately from the electric and gas distribution utility. With 1,300 workers, the electric transmission company would operate PG&E's 18,500 miles of high-voltage power lines. The gas transmission company would employ 750 people and run 6,300 miles of natural gas pipelines in Northern and Central California. The generating arm of the company would operate hydroelectric dams and the Diablo Canyon nuclear power plant, which together produce about 7,100 MW of electricity (about one-third of PG&E Co.'s total generation supplies). The generating arm would enter a 12-year contract with PG&E Co. to sell the energy from those plants for an average cost of five cents/kWh.

What PG&E Corp. is attempting to do is a classic example of a financial strategy known as ring fencing. In general terms, ring fencing includes efforts by a corporation to protect certain assets or liabilities by either creating a new subsidiary or cutting off corporate financing to an existing subsidiary. In this particular application, one of the most important aspects of the reorganization is that it would allow PG&E to increase its opportunities for financial backing, which are virtually non-existent now that the company has gone into Chapter 11 bankruptcy protection.

According to PG&E Corp., the reorganization plan would allow the utility to borrow more against its assets. One of the current challenges for PG&E Corp. is that California state law limits the amount a utility can borrow. However, by moving the utility's transmission and generating assets into new subsidiaries, the new companies could borrow against the full value of those assets. It is PG&E's hope that the reorganization would help it to raise money to pay creditors $9.1 billion in cash and $4.1 billion in long-term loans. All creditors owed less than $100,000 would get cash payments for the full amount once the plan becomes effective, which PG&E said could be by the end of 2002. PG&E Co., the utility operation, would become a separate company with its own stock, under the reorganization plan. Keep in mind also that PG&E Corp. had previously established a ring fence approach to protect its National Energy Group division earlier in 2001.

The current reorganization plan could violate California state law that prohibits such asset transfers without state approval, which is one of the key concerns of the plan's critics. But PG&E Chairman Robert Glynn said his company believes the plan is legal because "a federal bankruptcy judge has the authority to overrule state law" if it is in the interest of creditors and the company. Mr. Glynn added that he doesn't believe the plan is subject to review by the CPUC or other state officials.

Nevertheless, state officials have weighed in with their disapproval of PG&E Corp.'s restructuring plan. For one, California Gov. Gray Davis has stated that the plan gives the federal government (namely, FERC) too much control over the assets of PG&E Corp. (and, by the same token, reduces the amount of control that the state of California would have over these assets). Stating his position very clearly, Gov. Gray Davis said, "I am very wary of PG&E's proposal to transfer all of its generating capacity from a regulated environment to a non-regulated environment, which shifts oversight from the CPUC to the FERC." Further, Davis said that FERC has treated California ratepayers "shabbily" over the past 18 months and he believes that the CPUC should remain in control of PG&E Corp.'s assets to ensure a better deal for California ratepayers. In response to this criticism, PG&E Corp. says that its reorganization plan maintains the current regulatory authority for virtually all aspects of its business, and the CPUC will continue to regulate PG&E Co., including retail electric and natural gas rates.

Consumer groups also remain guarded about the reorganization plan. The Utility Reform Network (TURN) pointed out that PG&E Co.'s utility assets would be transferred to the parent organization. This would give PG&E Corp. the freedom "to gouge consumers without regulatory oversight," in the words of TURN's Executive Director Nettie Hoge. "Rates would certainly go up, because after snatching the assets PG&E Corp. would sell us (consumers) back the electricity at inflated rates," said Hoge. (From my perspective, this does not appear to be a true statement as the CPUC would clearly maintain its jurisdiction over retail rates.)

In response, PG&E Corp. claims that it will sell electricity back to consumers at market rates. However, market rates are still far above the cost of service rates required by the CPUC. According to TURN, consumers could end up paying about five cents/kWh for electricity instead of the cost of service rate of 2.8 cents/kWh, calculated for the hydro and nuclear generating assets. Even if this were true, the five cents/kWh would be on par with the prices charged by the Department of Water Resources for its current long-term power contracts.

One of the values for consumers and the state of California, according to PG&E, is that the restructuring would enable all valid creditor claims to be paid without the need for a rate increase or state bailout. My question for PG&E Corp. would be how long are they willing to commit to the "no rate increase" pledge. Will there be no rate increase for a month, six months, a year? One source of concern that appears valid for consumer groups is that PG&E has not committed to any timeframe in which rates would not be increased.

Loretta Lynch, president of the CPUC, says the plan is no more than a "regulatory jailbreak" that only serves the purpose of allowing PG&E to transfer valuable assets from regulated affiliates into unregulated affiliates. This is an important point as the state of California, under the leadership of Gov. Gray Davis, has increased the role that the state government is playing in the energy market. However, this criticism may be a bit overblown by the CPUC and consumer groups. If successful, PG&E's plan would only transfer the regulatory control over the generating assets themselves (including potential sales) from the CPUC to FERC, but the CPUC would still retain regulatory control over the retail rates for power from those generating assets and also maintain the right to review how PG&E Co. is buying power.

Further, from an investment perspective, PG&E Corp. says that its plan would allow the utility unit of PG&E Co. to pay off its debts in full with interest and make it solvent again, without a rate increase or state bailout. In addition, perhaps the bottom line regarding this reorganization plan is that PG&E Corp. claims that it can choose between only three alternatives to raise enough money to pay off existing debts: 1) raise rates or get a state bailout; 2) sell off assets to out-of-state generators; or 3) reorganize and refinance its existing assets. The corporation has chosen to follow the third option, which it views as the all-around best option.

However, as noted, the resistance to the plan (especially from the CPUC) may remain a major obstacle for PG&E Corp. The CPUC and other critics of the plan have asked U.S Bankruptcy Judge Dennis Montali to delay allowing PG&E to file further reorganization documents until it is fully clear if the company must gain state approval before proceeding. At this point, PG&E needs Judge Montali's approval to preempt state law and proceed with its reorganization. Without this approval, the plan could be significantly thwarted, as PG&E may be forced to enter negotiations once again with the state of California and the CPUC.

Ironically, PG&E Corp. announced in early November 2001 that its third-quarter net income had more than tripled. The company's net income rose to $771 million, or $2.12 a share, from $225 million, or 62 cents a share, in the third quarter of 2000. Excluding special items, PG&E Corp.'s earnings from operations were $256 million, or 70 cents a share, compared with $248 million, or 68 cents a share, a year earlier. PG&E Co. contributed operating income of $192 million, or $0.53 per diluted share, compared with $211 million, or $0.58 per diluted share, last year. PG&E National Energy Group grew operating results to $77 million, or $0.21 per diluted share, for the quarter, compared with $37 million, or $0.10 per diluted share, in the third quarter of 2000.

Southern California Edison

As of early fall 2001, the jury was still out regarding whether or not SCE would follow PG&E Co. into bankruptcy court. The prognosis did not look good for the utility, at least from a financial perspective. However, executives with SCE continued to hammer out a controversial rescue plan with Gov. Gray Davis and the California Legislature that would have the ultimate goal of keeping the utility financially solvent.

The steps leading up to what appeared to an approval of some sort of agreement between SCE and the State of California are important to track. Edison International posted a second-quarter loss on July 20, 2001, and warned that its SCE subsidiary would go bankrupt without immediate action from California lawmakers. "Simply put, SCE cannot avoid bankruptcy and the state cannot get out of the energy-buying business without immediate legislative intervention," Edison Senior Vice President Bob Foster said in a letter to state lawmakers. SCE's road to bankruptcy was essentially the same one that PG&E Co. followed. SCE accumulated billions of dollars of debt buying power at skyrocketing prices in the wholesale market. The costs could not be fully passed on to customers due to a retail price freeze imposed under the state's power deregulation laws.

On July 21, 2001, the day after Edison International posted its 2Q financial loss and issued the dire warning about SCE's potential bankruptcy, the California Senate passed what was attempted to be a rescue plan for the struggling utility. In exchange for allowing SCE to issue bonds to pay off its debt, the state of California would extend the timeline in which it may be able to purchase SCE's transmission assets. However, SCE quickly responded that the Senate's measure is a case of "too little, too late" and that even though many have claimed that the measure was an unwarranted bailout, it still did not offer an adequate resolution to the utility's financial problems. In addition, lawmakers continued to resist approving any measure that might be perceived as a financial bailout for SCE.

Under a 22-17 vote, the California Senate approved a modified version of an agreement between SCE and Gov. Gray Davis that had been in the works since April 2001. Specifically, the bill allowed SCE to issue up to $2.5 billion in bonds to be repaid by a mandatory charge on some customer bills. The money would be used to repay about $1.2 billion that SCE owes to financial institutions, and another $1.3 billion to qualifying facilities (smaller power plants that had provided power to the utility as a backup when larger suppliers began to grow concerned about the utility's creditworthiness). In addition, the bill would provide the state of California with a five-year option to purchase SCE's 12,000-mile transmission assets at a book value of about $1.2 billion.

There were several points about this bill that are important to note. First, SCE was offered much less (about $1 billion, in fact) than it said it needed to stave off bankruptcy proceedings. Second, whereas previously Gov. Gray Davis had been very eager to forge a deal to immediately purchase SCE's transmission assets, this aspect of the deal appeared to have cooled a bit. Under the revised agreement, the state still retained the option of purchasing the assets, but would have five years to finalize a decision. Also note that, in contrast to the "agreement in principle" that had been reached between Gov. Gray Davis and SCE back in April 2001, the state was now offering book value for the transmission assets, as opposed to the $2.76 billion that Davis had offered three months prior. At the time, legislators throughout California had balked at the high price for the transmission assets, which they believed were overvalued. In addition, some lawmakers thought that the high price that Davis offered for the transmission assets merely constituted a bailout for SCE, which gave a good number of legislators heartburn. In many ways the agreement in principle between SCE and Gov. Gray Davis represented a sweetheart deal for the utility as it would have been paid 2.3 times above book value for its transmission grid, which never has represented a strong revenue stream for the utility. By the July 2001 measure, the offering for the assets had been cut in half, and SCE clearly was no longer as pleased with the deal.

The bill also required that SCE pay off as much of its existing debt as possible and return to creditworthiness by January 1, 2003. At the time, and still as of this writing, SCE falls under a deep-junk credit rating and consequently has been forced to turn the bulk of its power purchases over to the state of California. Obviously, a financially sound utility benefits the state's economy, so all of the stakeholders in this matter appeared to be in agreement in wanting SCE to once again resume buying power for the customers it serves. The prerequisite for this, however, was getting SCE back into a financially solvent position, the method for which is something that has caused great discord among California legislators.

The other pressure on this deal making was time, and the clock kept ticking away. For starters, the agreement in principle reached between SCE and Gov. Gray Davis had a deadline of August 15. Unless an acceptable deal was reached between the two parties by that date, all bets were off. Compounding the problem was that California lawmakers began a 30-day recess as of July 20. The lawmakers continued to attempt to reach a compromise during this break, but realized that it would be very difficult to gain support for the rescue plan in the California Assembly. In fact, even though the Senate bill included a financial offering that was significantly less than Davis' original offer, some California legislators still considered the bill to be "corporate welfare" for SCE and vowed to block voting on the measure once it reached the Assembly.

All of which still continued to leave SCE in a very precarious state. SCE representatives warned that the Senate plan was inadequate and would leave the utility about $1 billion short of the amount it needed to become creditworthy again. Although the measure might have enabled SCE to pay some of the smaller generators and lenders that it owes, outstanding bills that the utility owes to larger power suppliers would remain intact. Thus, SCE representatives did not appear to support the Senate-approved bill because it did not fully restore the utility to creditworthy status.

Further, the staggering amount of SCE's debt was an issue that remained unresolved. As of the summer of 2001, SCE had reportedly accrued about $3 billion in debt, largely from previous purchases of power from qualifying facilities and other loans from lenders. The state of California, which took over the role of buying power for the state's utilities in January 2001 as a result of concerns about the utilities' creditworthiness, reportedly spent $27 billion on power purchases in 2000 and had spent $20 billion in the first six months of 2001. Some of the money to finance the state's power purchases had come from its existing general funds, which eventually had to be repaid. However, in order to pay off the balance and keep serving as a power broker, the state needed to receive additional financing.

One of the main stumbling blocks in the debate over SCE's financial future was the amount of financing that SCE and the state of California could secure, and how quickly the financial support would materialize. Convincing investors that the state of California and SCE made for sound investments would be a difficult task. For starters, the team led by State Treasurer Phil Angelides would have to convince bond-rating agencies such as Standard & Poor's and Moody's that the bond or bonds were secure and warranted a top-tier investment-bond rating. These firms had been burned by the California crisis already and would remain overly cautious about the soundness of any bonds issued. For instance, as noted in early 2001, Standard & Poor's had assigned investment-grade ratings to PG&E Co. three months before the utility declared bankruptcy. As the bankruptcy potential for SCE remains high, such investment-grade firms will not be apt to make the same mistake twice.

Further, any bonds issued to relieve the financial strain on California and SCE, which would have maturities of as long as 15 years, would be paid off by electricity ratepayers. This concerned investors for two reasons. First, capital garnered from customer bills can be rather fluid in nature depending on the possible decrease of the utility's customer base. In addition, and perhaps more importantly, retail rates remain regulated by the CPUC, and investors feared that any return on investment they made into the possible bonds could be restricted by unforeseen regulatory measures.

Investors also recognized the fact that many power generators were still owed billions of dollars for power that had already been sold either to the utilities or the state of California. Under the tenets of long-term contracts signed by the state, power generators typically would be paid first when SCE and the state recouped the bonds from customer rates. This also cast doubt on the timeframe in which investors might be paid for their own financial contributions. According to a report in *The Wall Street Journal*, all of these factors heightened ambivalence about the California bonds among investors, including groups represented by such capital firms as Envision Capital, Montgomery Asset Management, and Prudential Securities.

Moreover, the stakes surrounding the pending bond issue were high. On July 31, 2001, Edison International announced that it would not be able to raise funds to help pay off SCE's debts in the event that a state-directed rescue plan failed. California lawmakers had called upon Edison International to contribute money to assist in the resuscitation of its utility. However, the company's chief financial officer disclosed that Edison International had just been through an "extremely difficult refinancing process that was very expensive" and that the company's own financial status was precarious and linked to its subsidiary's possible bankruptcy.

In a surprise development in early October 2001, the CPUC and SCE announced that they had reached an agreement that would enable the utility to "pay off its creditors and prevent another utility bankruptcy in California." The CPUC also rejected the state's plan to issue a record $12.5 billion bond to fund some of California's power purchases.

The key elements of the plan included a rate freeze and an agreement by SCE not to pay shareholders a dividend until the debt is paid off. SCE's rates were raised by approximately 42% in early 2001 and will remain frozen through 2003 unless it pays off its debts sooner. In exchange, SCE agreed it would use cash on hand and any revenue beyond what it needs to cover operating expenses to pay off its old debts; pay no dividends on its common stock through 2003 or until its back debts are fully paid; and drop a lawsuit against state regulators claiming the CPUC had violated federal law by failing to raise retail rates to reflect the underlying cost of wholesale power.

Consumer groups blasted the settlement, calling it a bailout that froze SCE's electric prices at artificially high levels. Mike Florio, senior attorney for TURN, argued that the settlement actually has "unhealthy implications for consumers." While the state legislature "refused to force consumers to pay for the bailout they were considering, the CPUC is insisting that small customers bear the brunt of Edison's problems," he said, noting that consumers will pay inflated rates indefinitely.

Gov. Gray Davis was also supportive of the agreement stating it "protected the public interest and will allow the state's second-largest utility to return to financial health," adding that he welcomed the CPUC's assurance this could be accomplished without raising electricity rates. The governor canceled a special October 9 legislative session he had called to solve SCE's financial problems.

The governor, however, had harsh words concerning an earlier vote by the CPUC that rejected the state's bond plan to finance electricity purchases made by the state's Department of Water Resources on behalf of utilities. California's bond plan was defeated in a 4-1 vote, driven by concerns that the plan was too costly and would surrender the commission's authority to regulate power prices. The agreement would have allowed the state's Department of Water Resources to decide the "reasonableness" of power rates, traditionally a CPUC responsibility. Commissioner Henry Duque said this would "set a dangerous precedent" and lead to the "dismantling of the CPUC." "The Department of Water Resources was not meant to be a bloated power bureaucracy" and the rate agreement would do this, said Commissioner Richard Bilas.

The state of California must now go back to the drawing board to find a way to sell the record $12.5 billion bond. State Treasurer Angelides said he could not sell the power bonds until the CPUC "takes the actions required to implement the law. The CPUC's refusal to take action means there is no schedule for the bond issuance and no plan in effect to repay the state's general fund." Angelides warned that the state faces a budget deficit of over $9 billion in the 2002-2003 fiscal year if the bonds are not sold before July 2002. Gov. Gray Davis said the vote "was an irresponsible act. It creates uncertainty about our ability to sell the bonds necessary to repay the general fund when California can least afford additional fiscal uncertainty."

The four commissioners voting against the bond plan all favor a bill that was authored by state Senate Leader John Burton, a San Francisco Democrat, on grounds that it is more flexible and less costly. But the bill, already passed by the legislature, is opposed by Gov. Gray Davis, who has vowed to veto it when it lands on his desk.

Interestingly, the issue of SCE selling its transmission lines to the State of California now appears to be void. As discussed in chapter 7, Gov. Gray Davis had consistently maintained that there would be no value to the state in purchasing transmission lines of only one of the utilities in California, and that he was only interested in pursuing this plan if all three state IOUs agreed to sell their transmission assets. PG&E Co. had consistently maintained that the State of California was not willing to pay the utility what it thought its transmission assets were worth, and thus had rebuffed all attempts at negotiations by Gov. Gray Davis. This probably played a significant role in negotiations breaking down for the sale of SCE's transmission assets to the State.

Moreover, true to its word, SCE seemingly has found a way to avert bankruptcy. This is a major milestone in the restoration of California's power business. But challenges remain for the state. PG&E Co. remains in bankruptcy, the state is saddled with a $12.5 billion debt that it must find a way to finance, and wholesale electricity prices continue to decrease, making the state's decision to enter into very long-term contracts for electricity supply very questionable. Ironically, California began its deregulation efforts on the hopes of lowering electricity costs, which at the time were some of the highest in the nation. Now, the state is locked into even higher prices, as SCE's interim-rate increases averaging four cents/kWh are permanent for the foreseeable future.

Chapter 5
Resolution Attempts

One of the central challenges associated with resolving the California energy crisis was the uncertainty over jurisdictional responsibility. For months, as the crisis continued to worsen, California officials such as Gov. Gray Davis and the members of the California Legislature continued to criticize the FERC and the administration under President George W. Bush for not intervening with measures to resolve the crisis. Typically, the measure that state officials most often requested was the implementation of price caps for the sale of wholesale power in California and the entire Western region. In turn, FERC commissioners, President Bush, and Vice President Dick Cheney maintained that the California energy crisis was a state issue and needed to be solved through state measures. Regarding the issue of price caps specifically, federal officials in general continued to espouse the counter-argument that price caps would not solve the problems plaguing California's energy market and in fact could cause more harm than good. This debate over jurisdictional responsibility undoubtedly resulted in the California energy crisis lingering far longer than was necessary.

Somewhere in mid-2001, the California energy crisis began to subside, due largely to a combination of factors including mild weather patterns, increased conservation efforts, and the fact that new generation sources had quickly become available. However, the dissipation of the California energy crisis was also due to the fact that a coalescence of state and federal resolution efforts began to take effect. In this chapter, we will examine the various steps taken by state officials and federal regulators that, when taken together, helped greatly to break the devastating electric crisis that had occurred in California.

STATE MEASURES

Rate Increases

As noted in chapter 4, one of the first ways in which the California energy crisis manifested in a real-world way were the price spikes that occurred in San Diego after SDG&E paid off its stranded costs, which ended the rate freeze in the area. In September 2000, the California Legislature approved a pair of measures to use $150 million in taxpayer money to help San Diegans with the electricity price spikes and to immediately cut electricity bills in San Diego. AB 265, as the legislation was called, proposed some significant measures. First, $150 million collected from California taxpayers would be set aside until 2003. If at that time, it is determined that SDG&E has encountered financial losses so severe that to cover the losses would mean an additional 10% increase on customer bills, this reserve fund would be used to compensate SDG&E customers. In addition, the bill called for an immediate cut in current SDG&E rates—which had nearly doubled or tripled in the summer of 2000—for all but the largest customers of the utility. Reports indicated that the average residential bill in San Diego would drop from $128 to $68 as a result of the legislative action.

Within this emergency measure was a clause that put a rate cap of 6.5 cents/kWh on SDG&E rates for residential, small commercial, and lighting customers. The bill mandated the CPUC to initiate a voluntary program for large commercial, agricultural, and industrial customers of SDG&E to also set the energy component of their bills at 6.5 cents/kWh with a true-up after one year. That was the justification for having taxpayers create the $150 million fund. The rate freeze would be in effect until June 1, 2002. Should SDG&E need to raise rates after that date to recover losses due to the rate cut, SDG&E customers would be further protected by the safety net measure.

The measure, penned by Assemblywoman Denise Ducheny (D-San Diego) and Assemblyman Jim Battin (R-La Quinta), was passed by the Assembly 58-12 after previously clearing the California Senate.

Reaction to the passage of the measure in the state legislature was swift and typically heated. Sen. Tom Hayden joined a group of dissenting Republicans and noted that the bill constituted a "bailout" for SDG&E.

The passage of the bill came after months of horrible press for California and demands from outraged customers in San Diego that the state and federal government intervene. As discussed previously, starting in the spring of 2000, SDG&E became the first IOU to pay off its stranded costs, thus ending the mandated rate freeze that protected its customers from wholesale market prices of electricity. SDG&E was able to pay off its stranded costs earlier than the March 31, 2002, deadline due to the sale of its two fossil-fuel plants at greater than book value. Fellow California IOUs, SCE and PG&E Co. were still paying off their stranded costs, and thus their customers remained protected under a rate freeze.

Due to a number of factors—including power supply shortages within the state, increased demand, and sweltering heat—electricity price spikes occurred that drove up the cost of SDG&E bills dramatically. In early August, the CPUC approved a $390 million payback for SDG&E customers to compensate them for the increase in bills. The legislative action of AB 265 was viewed as an additional support mechanism that will now come directly out of taxpayers' pockets.

On September 8, 2000, Gov. Gray Davis approved emergency legislation to cut electricity rates in San Diego, but did not sign the measure that would have established the $150 million fund, to be paid by taxpayers, to compensate energy customers. The law signed by Davis puts a rate cap of 6.5 cents/kWh on SDG&E rates for residential, small commercial, and lighting customers, retroactive to June 1, 2000. This rate cap was less than a third of the then "free market" cost of 21.4 cents/kWh.

The law also mandated the CPUC to initiate a voluntary program for large commercial, agricultural, and industrial customers of SDG&E to also set the energy component of their bills at 6.5 cents/kWh with a true-up after one year. The rate freeze will be in effect until June 1, 2002, although the option exists to extend it through December 2003 if the CPUC finds that it is in the public interest.

SDG&E, which had urged Davis to veto the legislation, responded by saying that the law was a "quick political fix" which would do nothing to repair severe structure problems in California's wholesale power market. SDG&E estimated that the under-collected shortfall it would bear because of the rate cap would amount to $664 million by the end of 2002. If the rate cap were extended for another year, SDG&E would lose around $839 million. Consequently, although this law by Governor Davis provided some immediate relief for SDG&E customers, the continuance of what appeared to be manipulation of market transactions signaled a much larger issue for California that had yet to be resolved.

Upon SDG&E receiving a rate increase, attention turned to PG&E Co. and SCE and the question of whether or not state regulators would also provide rate relief to these two utilities, which continued to accumulate staggering debts. Three months later, on December 21, 2000, the CPUC agreed to give PG&E Co. and SCE a rate increase to stave off potential bankruptcy. The rate increase, the amount of which was undisclosed, took effect in January 2001. The CPUC also ruled that independent auditors must first examine the utilities' books before the rate increase could take effect. The CPUC said that the rate increase was necessary to protect the credit worthiness of both PG&E Co. and SCE from "sinking into a sea of debt that has been run up this year by the soaring cost of electricity purchased from private generating companies throughout the West."

Although this was mostly good news for both PG&E Co. and SCE, because they had an agreement from the CPUC to raise rates, it is important to note that no specific amount of the increase had been determined. This had been, and will continue to be, the main sticking point of ongoing negotiations between the utilities and the CPUC, as the utilities have argued that they should not be forced to absorb any portion of their massive debt. The inspection of the utilities' books is also an important factor. The purpose of the inspection was to determine if the utilities' financial situations were as grave as they had claimed and whether or not raising rates was absolutely necessary.

The CPUC's ruling was an "interim decision," indicating that it recognized the severity of the problems in California and would take definitive action. Within this interim order, the CPUC agreed in theory to a rate increase of an unspecified amount, contingent upon the inspection of PG&E Co. and SCE books. The CPUC acknowledged that action must be taken "to ensure that the utilities can provide service at just and reasonable rates" and that any action that "may jeopardize the utilities' creditworthiness and their ability to continue to procure energy on behalf of customers" must be avoided. The CPUC also agreed that retail rates must increase in the state. SCE reacted to the interim order by saying that it wished the CPUC had acted more decisively, but that it would continue to work with the commission to reach a final order. In addition, SCE said that the crisis required "prompt, meaningful, and final action by the commission to end the rate freeze, provide for the recovery of procurement costs, and increase rates to a level that will restore our creditworthiness."

As the CPUC continued to formulate a decisive order, rumor had it that state officials would propose an increase in the amount of 10%—which would take effect after an immediate lifting of the existing rate freeze—and force the utilities to swallow a certain portion of the debt they currently claim. PG&E Co. and SCE had repeatedly demanded rate increases of almost double this offer, and wanted to minimize the amount of debt that they would have to swallow.

Nevertheless, the CPUC already had agreed in theory to a rate increase, but the sticking point was the specific amount of that increase. Reportedly, the CPUC and Gov. Gray Davis proposed a 15% rate hike to the two utilities. In sharp contrast, PG&E Co. maintained that it needed to increase rates by as much as 40%, and SCE said that it wanted an increase of 30% (or 76% over two years). Both utilities continued to argue that even if their proposed rate increases were accepted, they still could not be enough to cover wholesale costs, which were projected to rise through 2003. In addition, PG&E Co. and SCE also claimed that their financial solvency still would not be guaranteed even if their rate increase proposals were accepted, and bankruptcy would be very likely if the CPUC only allowed them to increase rates by 15%.

It is important not to overlook or discount the role that consumer advocates played in the negotiations. Harvey Rosenfield, a long-time critic of the state's utilities who led the failed attempt to overturn deregulation in California in 1998, remained very vocal that "the people of California should not have to pay one more penny to bail out the utilities for the mistakes they made." Consumer groups led by Rosenfield and Ralph Nader continued to put enormous pressure on Gov. Gray Davis, a Democrat, to keep any rate increase given to the utilities to a minimum. In fact, Nader stated that Davis' political future "hangs by a kilowatt hour" and that any decisions made in California would "reverberate all over the country." In essence, consumer groups believed that utilities shared the bulk of responsibility for the economic mess in the state, as they played a large role in California's 1996 restructuring legislation. If mistakes were made in that legislation, consumer groups said, the utilities and not utility customers should be made to pay the price.

In turn, PG&E Co. and SCE argued that the real cause of the problem in the state was the generators, which unjustifiably continued to charge exorbitant rates for power on the wholesale market. There may be some validity to this theory as power generators that operated in California knew that the state's IOUs had essentially been stripped of their own ability to generate power and thus were heavily dependent on power bought on the wholesale market. This gave an enormous amount of market power to the generators and allowed them to charge rates that were not reflective of their own costs.

On January 4, 2001, the CPUC recommended energy rate increases for the state's two cash-strapped utilities, but gave the companies far less than they wanted. The CPUC tentatively granted PG&E Co. and SCE surcharges that would result in rate increases of about 9% for residential customers, or about $5 a month on the average bill. The increases were not as large as what the utilities asked for, and would last only 90 days.

To say that the rate increases were not as large as what PG&E Co. and SCE had asked for (demanded, really) is indeed an understatement. PG&E Co. had requested an immediate 26% increase and SCE had asked for 30%.

Thus, the amount proposed by the CPUC was less than half of what both utilities indicated was the necessary rate increase to stave off bankruptcy measures. Both companies argued that they were on the brink of financial insolvency and should have the right to bill customers to compensate for the approximately $9 billion in debt that still hung over both companies.

In addition to the lower-than-expected rate increase, the proposal from the CPUC came with an important catch. PG&E Co. and SCE would not have immediate access to the money generated by the rate hikes. Instead, any accrued funds would be put into an account, controlled by the CPUC, until an independent audit determined that the utilities did in fact need the money. If the audit determined that the utilities did not need the money, it would be returned to the utilities' customers. This aspect of the proposal indicated continuing speculation that both PG&E Co. and SCE were cooking their books to make their financial positions appear far graver than they actually were.

As expected, the minimal and temporary rate increases did little to help the financial positions of either PG&E Co. or SCE, and both utilities continued to issue threats that bankruptcy would be imminent without substantial and long-term rate increases. Consequently, in March 2001, the CPUC approved electricity rate increases topping 40%, the largest in the state's history. Specifically, the CPUC unanimously decided to raise rates by three cents/kWh. The rate hikes, which took effect immediately but did not appear on customer bills until May 2001, applied to millions of customers of Southern California Edison and PG&E Co. Low-income residents were exempt from the increases.

For context, keep in mind that it was only two months earlier that the CPUC approved its first rate increase in response to the staggering debt of SCE and PG&E Co. and other factors related to the California energy crisis. In early January, the CPUC approved temporary rate hikes of 9% for residential customers (about $5 a month for the average customer) and 7–15% for businesses. The temporary increases were only to remain in effect for 90 days. At that time, both PG&E Co. and SCE denounced the temporary increase, which was not surprising considering that the 9–15% increase was much lower than the utilities had requested. The January rate increase was followed by an additional 10% increase.

The CPUC's new ruling increased rates by about 42% at SCE and 46% at PG&E Co. However, the CPUC's ruling most likely benefited the state more than SCE or PG&E Co. The CPUC clearly established that none of the additional revenue earned from the rate increases could be used by the utilities to pay for past power purchase costs, which had amounted to about $13 billion. In addition, the CPUC adopted a proposal by the consumer group TURN, that a rate cap preventing the utilities from charging market rates should be extended. In other words, the portion of the electric rates that PG&E Co. and SCE actually received from ratepayers would remain frozen for the foreseeable future. The CPUC established that the utilities must still recover billions of dollars in stranded costs before the rate freeze could be lifted.

Thus, it appeared that most of the revenue accumulated from the rate increases would be used to reimburse the state of California for the power purchases it continued to make on behalf of the utilities and would not be used to pay off any existing debt held by the utilities. PG&E Co. responded that the CPUC ruling attempts to "completely change the ratemaking rules that are used to determine the end of the rate freeze under AB 1890." PG&E Co. officials said that it would challenge this and other aspects of the CPUC ruling.

However, even though most of the money from the rate increases would go to the state, it was also not clear if the increases would be sufficient to reimburse California for the $4 billion it has already spent to procure power, or compensate for the anticipated $14 billion that it will have to pay for power over the course of 2001. Some of the funds generated by the rate increases must be used by PG&E Co. and SCE to pay for power that they still obtained from suppliers or to cover costs for their own power generation. Consequently, the CPUC has left open the possibility that additional rate increases may be necessary.

Although it was uncertain how individual customer classes would be impacted by the rate increases, it was fairly clear that large commercial and industrial customers would end up paying the most. Loretta Lynch, president of the CPUC, maintained that an underlying goal of the increase was to force what she referred to as "electricity hogs," or those people who excessively use power without concern for the state's supply shortage, to conserve. Lynch said that conservation was the key to staving off additional rate hikes, and that the higher rates should put financial pressure on Californians to conserve energy, which theoretically should help to reduce the risk of blackouts in the state in the summer of 2001. The average residential cost for a kilowatt hour at that time was about 12.5 cents for SCE customers and 10.5 cents for PG&E Co. customers.

However, Lynch also established that a "tiered" pricing structure would be put into place. The tiered pricing system would reward customers who conserved energy and penalize those who do not. Essentially this meant that those customers who use the most power would ultimately pay the most, and could face additional rate increases this year, which the CPUC did not rule out as a possibility. Under a tiered pricing structure, those customers who conserved on their power usage might not see their bills increase beyond the rate increase.

Gray Davis, who throughout his tenure as California's governor assured Californians that rates would not be increased, attempted to distance himself from the CPUC's ruling, and in fact issued a statement that he had no control over the commission's decision. Further, Davis said that the rate increase was "premature" because the state was still evaluating the financial numbers that were needed to make a sound decision. In turn, the CPUC's Lynch criticized both the governor and the FERC for "failing to act" on energy rates up to that point.

In addition to its comments regarding the rate freeze, PG&E Co. responded with a mixed reaction to the CPUC's ruling on the rate increase. On one hand, the utility said the ruling offered a "welcome dose of realism." However, the utility also said that the rate increase "does not offer a comprehensive solution, fails to resolve the uncertainty of the crisis, and may even create more instability." Based on previous statements that PG&E Co. has made, the utility believes that the state's fundamental problem has its origins in the wholesale market, where prices have often exceeded $300/MWh.

Some power generators, many of whom have been lambasted by California and federal regulators for alleged price gouging, went on record with a favorable response to the rate increases. Jeffrey Skilling, who at the time was Enron's CEO, said, "There's been a need to balance supply and demand [in California] and to date I don't think anyone has been willing to let the prices rise to help balance supply and demand. So [the increases] are a huge step forward in reducing the odds of shortfalls." Keith Bailey, chairman and CEO of Williams, told Reuters that he thought the CPUC's rate increase was a rational decision and something that the industry has been suggesting for some time.

Naturally, many consumer activists reacted vehemently against the rate increases and protested during the CPUC's proceedings. The main argument of the consumer groups was that the CPUC, the governor, and FERC had not fulfilled their regulatory responsibilities in bringing down electricity prices in the state and have not acted with enough force to restrict what they perceived as excessive prices for power owned by out-of-state generators.

Moreover, while the rate increases apparently helped the state to continue securing power on behalf of the utilities, other fundamental problems related to the state's energy crisis remained unresolved. The CPUC asserted that part of the benefit of the rate increases was to keep the utilities financially solvent. However, at the same time, the commission clearly stated that none of the extra revenue generated by the rate hikes could be used to pay down the utilities' outstanding debt. Consequently, although the higher rates should significantly increase revenues for both utilities ($2.3 billion a year for SCE and $2.5 billion a year for PG&E), all of this money would be turned over to the state and do little to alleviate the utilities' own financial problems.

Governor Davis' Resolution Plan

In addition to separate resolution efforts being taken by the CPUC, Gov. Gray Davis also implemented his own extensive measures to bring stability to the electricity market in the state. In his State of the State address delivered on January 8, 2001, Gov. Gray Davis said electricity deregulation in the state had been "a colossal and dangerous failure" that had resulted in "price gouging." Further, the governor said that "never again can we allow out-of-state profiteers to hold Californians hostage, or allow out-of-state generators to threaten to turn off our lights with the flip of a switch." The governor proceeded to outline a controversial plan to resolve the energy crisis in California, a plan that ranged from altering the way in which the state's Power Exchange operated to dedicating $1 billion to fund the construction of new power plants.

In what was generally considered the most critical speech of his career, Gov. Gray Davis did go out on a limb to propose several resolutions to ongoing problems in California. First, the governor's preference for the state assuming greater control over its own energy market became clear in his address. Specifically, Davis said, "the time has come to take control of our own energy destiny." What this would mean in real terms is that the state government would establish a state power authority to build new power plants. The governor also vowed to overhaul the "crazy bidding process" of the state's energy auction (the California Power Exchange), in which each generator is paid according to the highest bid rather than their own bid. In addition, the governor proposed expanding his own emergency powers "in the event of imminent power outages."

Davis repeated an earlier pledge to set aside $1 billion in his budget to help bring new power plants online and encourage energy conservation. Regarding conservation, Davis asked all Californians for a voluntary 7% reduction in power consumption. As has been previously documented, no new major power plant has been constructed in California in almost two decades. Davis called for "cutting the red tape" that had kept new power plants from being built in California and for committing public lands to that purpose "on the condition that energy [from those plants] be distributed in California."

Davis' harshest words were levied at power suppliers. The governor clearly accused power generators of price gouging (or improperly inflating the amount they are charging utility buyers for power on the wholesale market) and gave the state attorney general $4 million to investigate and prosecute market-manipulation cases. Davis even proposed making it a criminal act punishable by a prison term for generating companies to withhold electricity.

Ironically, in his speech Davis made no specific proposal for helping PG&E Co. and SCE to extricate themselves from the current $11 billion in debt that they faced as of January 2001.

In follow-up measures to his State of the State address, Calif. Gov. Gray Davis and legislative leaders agreed on a plan to resolve California's electricity crisis that included years of power-buying by the state. "The state will be in the power business for a long time to come," Davis said. He said details were still being worked out, but would include state ownership of utilities and long-term electricity purchases for the cash-starved companies.

It had been clear since the governor's State of the State address that Davis believed the state government should (or must) play a larger role in California's energy market. Toward that end, the California Department of Water Resources began to establish long-term contracts to procure power on behalf of the utilities, using set-aside emergency funds of $400 million. Reportedly, all that money was quickly spent on power purchases, but the governor ordered that other state funds be used to continue buying power for the utilities. Operating on a belief that it could secure better rates than debt-ridden PG&E Co. and SCE, the state intended to supply roughly one-third of California's supply needs. A 27-hour auction resulted in 39 bids from power suppliers with an average 6.9 cents/kWh offer. Although higher than Davis' goal of 5.5 cents/kWh, the average bid offer was closer to then-current retail rates and far below what the utilities themselves had been paying on the spot market. The Davis administration further committed $2.7 billion for power, about $45 million a day, as of March 2001. That amount would ultimately need to be repaid when the state issued revenue bonds, an issue that remains a contested and unresolved matter as of this writing.

A key condition of the new role that the state had assumed was that the Department of Water Resources would only pay for power at prices that it deemed "reasonable." Any remainder that might be necessary to meet the power needs of the incumbent IOUs would be left up to the California ISO to purchase in the real-time market.

On March 5, 2001, Gov. Gray Davis announced that the State of California—through the Department of Water Resources—had agreed to 40 power purchase contracts with more than 20 generators and marketers, which would supply an average of 8,886 MW per year over the next 10 years. The contracts, varying in length from four months to 10 years and worth $43.4 billion in total, reportedly will provide the state with a diversified long-term power portfolio, said Davis in an announcement.

According to a report in *Megawatt Daily*, the average price of the deals is $69/MWh over the entire 10-year period. For the first five years, the average price will be $79/MWh, dropping to $61/MWh for the period between 2006 and 2011. Gov. Gray Davis has offered that these prices are between 75 and 80% lower than recent spot market prices, although they are higher than the original goal of $55/MWh that Davis had established when he first began to negotiate long-term contracts.

There were a few public announcements regarding the long-term contracts, so we know some of the 20 companies that have entered into agreements with the state. For instance, Calpine Corp. locked the Department of Water Resources into a 10-year contract for the sale of approximately 5,000 MW of electricity. California will be paying a fixed price for the power it buys from Calpine, but most reports have placed the value of the deal in the range of $4.6 billion. Dynegy and NRG Energy said that they have agreed to sell up to 2,300 MW to the Department of Water Resources under contracts lasting through 2004. According to the companies, their California generating affiliates—El Segundo Power LLC, Long Beach Generation LLC, and Cabrillo I—will provide up to 1,000 MW of power to the state through 2001, increasing deliveries of 2,300 MW from January 1, 2002. Mirant Corp. (formerly Southern Energy) announced an agreement to shift 1,000 MW in power contracts from the California XC to the state's Department of Water Resources.

These are just a few examples, among other companies such as Avista, Duke, Enron, PacifiCorp, and Williams that signed on to serve the state under long-term contracts. Absent from the list was Reliant Energy, which reportedly engaged in negotiations with the Department of Water Resources, but was reluctant to commit to more than a 30-day contract to sell power to the state. Reliant is owed over $300 million for power delivered to Southern California Edison and PG&E Co. in November and December 2000. The two major utilities owe more than $12 billion to a handful of power-generating companies and can't buy more on their own, having no credit or cash to make the purchases, which as noted is what prompted the state to appoint the DWR into the power purchasing role in the first place.

Gov. Gray Davis has said, "With these deals in place, California's energy future is looking a whole lot brighter." But is it really? That is still the question. Representatives for the governor have indicated that the long-term contracts will reduce the state's need to purchase power on the spot market over the next 10 years by about 75%. Yet, of more immediate concern to analysts is what will transpire during the summer heat in 2002 and 2003, when demand increases. This could represent another precarious scenario for the state of California.

Expedited Process for New Generation

In addition to resolution steps outlined in his January 2001 State of the State address, Gov. Gray Davis also signed into law a bill creating a fast-track approach to building new power plants in California. The shortage of power supply within the state has been one of the factors cited as the cause for the price spikes in the San Diego market in the summer of 2001. The power plant law should help expedite the permitting process for the construction of new plants in the state, which on average have taken between seven to 10 years to become operational.

The problems in California unquestionably have been extreme—and arguably very isolated due to the unique structure of California's energy market. The state is known for its strict power plant siting and approval process, and potential generators both from within and outside California historically have faced stringent political and regulatory hurdles involved with getting a power plant approved. According to figures from the CPUC, from 1996 to 1999, California's peak demand increased a net of 5,522 MW, but capacity additions had netted only 672 MW (or 2% of the 55,500 MW online). Supply additions did not kept pace with demand increases, which is one of many causal factors that resulted in sky-high prices in San Diego.

To resolve this particular part of the larger crisis, Gov. Gray Davis said he intended to have 5,000 MW of new generating capacity available in California by July 2001, more than double the amount expected just a few months before. Davis said the new plants would provide enough electricity to ward off blackouts during that summer. Davis announced a series of orders intended to bring online 5,000 MW by July 2001, with an additional 5,000 MW by July 2002. This projection, aimed to provide some reassurances about the California crisis, comes at a time of legal hurdles for both California's state government and its major electric utilities.

In his series of executive orders, Gov. Gray Davis established a directive to get more plants running sooner and running harder in an attempt to avert any problems with warm temperatures. Gov. Gray Davis established that the first 5,000 MW he expects to have online by July 2001 would come from the following sources:

- Approximately 2,100 MW from new peaker units—which help the state through the busiest, or peak, hours of usage. The 2,100 MW would come from 30 "peaker plants"—basically big jet engines that generate roughly 50 MW of power each.
- 1,630 MW from renewable energy sources, such as wind and solar.
- 1,260 MW from large plants that are already under construction.

Those plants that met these criteria reportedly would be able to apply for permits from the CEC within 21 days, a process that normally has taken a year in the state. Plants that did in fact make it online by July 2001 would be eligible for bonuses of as much as $1 million.

Calpine Corp. is one company that stands to gain from Gov. Gray Davis' plan, but the relationship between the company and the state of California is clearly reciprocal. In mid-2001, Calpine had four plants under construction in and around California that totaled over 2,500 MW, four power plants in announced development totaling 2,900 MW, and an additional 3,700 MW that was in late-stage development. In late November 2001 Calpine stated that it had a program in place to build some 7,700 MW of clean, energy-efficient generation throughout California and in nearby states, representing a $4 billion investment in this region. Reportedly, by 2006, Calpine will have over 10,000 MW of low-cost generation in operation for California customers (or approximately 18% of anticipated demand).

In February 2001, Constellation Energy Group (NYSE: CEG) announced that Kiewit Industrial Company of Omaha, NE, would begin construction in April 2001 of the 750-MW High Desert power plant near Victorville, CA. The project, which had been in development since the mid-1990s by High Desert Power Project, LLC—a subsidiary of Constellation Energy Group of Baltimore, MD—would be the first new major power plant to serve the Southern California area in more than a decade.

The High Desert power plant will be built on a 25-acre site at the Southern California Logistics Airport, formerly known as the George Air Force Base. Commercial operation of the plant is scheduled for June 2003. The plant's electric generating technology is based on clean burning, natural gas-fired combustion turbines operating in a combined-cycle mode. Part of the supply problem in California was the fact that hydroelectric resources, one of the state's primary energy sources, had been in short supply due to lower-than-average rainfall over the last year. Moving away from a dependency on hydroelectric supply may be a smart move for the state. In addition, the High Desert plant is consistent with a national trend that supports natural gas as the fuel of choice for new power plant development projects. As a point of interest, according to data from the Energy Information Administration, California's fuel mix is primarily based on a mixture of natural gas (45.59%), hydro (21.95%), nuclear (18.07%), coal (1.36%), oil (1.04%), and other, presumably renewable, sources (11.98%).

The announcement of Constellation's new plant could not have come at a better time for California. Although the plant won't be operational until June 2003, the state needed the good publicity in early and mid-2001. A report from the Cambridge Energy Research Associates (CERA) released in February 2001 said that it was already too late for the state to head off a serious shortage in the summer of 2001 that could lead to at least 20 hours of rolling blackouts. "It's too late for this summer," declared Lawrence J. Makovich, of CERA, referring to the state's desperate efforts to ramp up power supplies for the short term. "The source of California's far-reaching power crisis is a shortage. At least 5,000 MW of new generating facilities are required to restore balance to the state's grid." Instead, according to the report, demand will exceed supply by 3,000 MW, which will cause the state to be under an energy alert of some sort for at least 200 hours.

According to Gov. Gray Davis' office, starting in April 1999, well before the California energy crisis began to receive front page coverage, the state of California began licensing new power plants. As of June 2001, Gov. Gray Davis claimed that the state had licensed 16 new plants, 10 of which are presently in construction. Four out of the 16 plants were expected to be online by the end of the summer of 2001. In addition, the state has reportedly approved 10 peaker plants (units that can be used only when demand runs exceedingly high to supplement baseload units). Gov. Gray Davis claimed that the peaker units were approved in less than 21 days due to his emergency power, which represented a reduction from the standard four months that were commonly necessary for approval. Looking forward, Gov. Gray Davis claimed that by the fall of 2003, California will finally have more power than its needs. Further, according to the governor, the state should have a 15% surplus of power by sometime in late 2004.

According to information from the California Energy Commission (current as of November 1, 2001), 29 total power plant projects have been approved since the start of deregulation in 1998, although not all of the plants will be built. Three "major" power plants, totaling 1,415 MW, have come on line in 2001 and are producing electricity. Another 864 MW from "peaking" power plants were scheduled to come on line by the end of September 2001. As of November 1, 2001, the California Energy Commission reported that 20 plants were under review, totaling 10,643 MW. These applications included simple and combined-cycle plants. Also as of that date, the California Energy Commission had approved four emergency peaker units, under the expedited 21-day approval schedule outlined by Gov. Gray Davis, which totaled 435 MW.

Additional and current information about the locations of California power plants approved by the California Energy Commission can be found at the Commission's Web site (www.energy.ca.gov).

FEDERAL EFFORTS

Resolutions Orders Issued by FERC

Along with state-led efforts, after increasing pressure from California officials and intense national media coverage of the energy crisis, federal officials (namely, the FERC) began to initiate their own resolution measures. In late 2000, FERC issued a draft order outlining a major overhaul of the California wholesale market, including changes to market rules concerning the California ISO (Cal-ISO) and PX. Conceding that the market rules and structure for wholesale sales of electricity were "flawed" and had caused "unjust and unreasonable" rates, FERC proposed a number of remedies.

FERC based its 77-page draft order on several assumptions, and through the order intended to offer long-term (rather than band-aid) solutions. First, FERC wanted to hold overall rates to competitive levels that would benefit consumers, while at the same time induce sufficient investment in capacity to ensure adequate service. Second, FERC acknowledged that some matters that had adversely impacted the California wholesale market were not under its jurisdiction, but that of the CPUC. Thus, FERC focused on matters within its "exclusive jurisdiction," even if some of the proposals preempt prior state decisions, while at the same time urging the state of California to rectify issues within its own jurisdiction. This is an important point, as FERC maintains that unless the CPUC addresses issues such as the inadequate siting of generation and transmission and demand-side response, California customers would still be exposed to higher prices that result from low supply. Third, FERC could find no evidence of market power abuses, although it did concede that the market in California certainly allows the opportunity for market abuse when energy supplies are tight.

Further, FERC contended that California IOUs have been bidding up to 80% of their load into the day-ahead market and hour-ahead spot markets, creating substantial short-term cost exposure. Due to regulatory restrictions from California's restructuring law (AB 1890), the IOUs were not able to pursue a balanced portfolio, including long-term and intermediate contracts. The central mission of FERC's order was to remove the restrictions that kept California IOUs from moving significant amounts of wholesale transactions into forward markets. The less reliant that the IOUs became on spot markets, so went FERC's theory, the less chance of price volatility—which should not only reduce costs for customers, but also increase reliability and increase prospects for new generation.

With that established, FERC's draft order offered the following resolution measures: 1. Eliminate the requirement that California's three IOUs (PG&E, SCE, and SDG&E) must sell into and buy from the PX. This proposal essentially permitted the three IOUs to establish bilateral contracts with energy suppliers, which was previously restricted under AB 1890. 2. Require market participants to schedule 95% of their transactions in the day-ahead markets. A penalty charge would be affixed for deviations in scheduling in excess of 5% of an entity's hourly load requirements. Disbursement of penalty revenues would be distributed to the loads that scheduled accurately. 3. The establishment of independent (non-stakeholder) governing boards for the California ISO and PX. 4. The establishment of generation interconnection procedures.

FERC also wanted the CPUC to address the following issues: delays in siting additions of generation and transmission capacity; implementation of additional demand response programs at the retail level; and elimination of impediments on load-serving entities pursuing power supplies on a forward basis.

"Soft" Price Cap

Perhaps the most contentious item within the order was FERC's decision to set a $150/MWh rate cap so that bids above this amount cannot set the market clearing price that would be paid to all bidders. This policy negated a previous proposal from the Cal-ISO, under which it lobbied for a $100/MWh bid cap (reduced from $250/MWh) on electricity purchases in the ISO's spot market. FERC's order froze the ISO bid cap at its current $250 level for the following 60 days. However, beyond that FERC instated what was being referred to as a "soft price cap" of $150/MWh, to remain in place until December 2002. It's referred to as "soft" because sellers may bid above this level and receive their bid if they are dispatched, but anything higher than $150/MWh will not set the price that all generators will receive. Also, any generator setting a bid above $150/MWh must report their bid to the Commission, and presumably fall under intense scrutiny.

The order revealed again some dissension with the FERC itself. Although generally the four commissioners that delivered the order (Hoecker, Massey, Breathitt, and Hébert) probably would agree that it is not perfect, then-Commissioner Hébert voiced the strongest disagreement with the order's principles. Hébert "hesitantly concurred" with the order, although he believed the commission went too far in its attempt to mitigate prices, something he believes FERC is ill equipped to do. Specifically, Hébert dissented with the decision to place any sort of price cap on wholesale transactions, preferring instead—as he advocated throughout his tenure as a FERC commissioner—"to entrust market participants with the ability and responsibility to mitigate their price exposure as they deem best."

Moreover, it was questionable whether or not FERC had achieved its goals of providing long-term solutions to California's problems. Rather, it seemed that the commission had sent a mixed message. On one hand, it allowed the California IOUs the freedom to buy and sell power outside of the PX, which seemed like a very pro-competition policy. Yet, on the other hand, FERC established a price cap (albeit a soft, not hard, one) on wholesale transactions, which arguably would continue to restrict true competitive market forces from materializing in California. The commission sent a clear signal that all bids should be under $150/MWh and would undergo interrogation if they were not, which essentially put political pressure on suppliers to maintain a price cap.

However, if one of the clear goals of this proposed order was to bring down energy prices for California customers, removing the requirement of the IOUs to buy and sell through the PX was a proactive step. The proposal changed the California structure to one in which local distribution companies (LDC) could possibly wield more market power than the energy suppliers, who essentially controlled prices when going through the PX. Of course, FERC's proposed order was implemented before the state of California took over the role of power purchaser through the Department of

Water Resources (taking away from the utilities the right to purchase their own power), but FERC's philosophical approach should be noted.

In December 2000, FERC unanimously approved the proposed order on ways to fix California's troubled marketplace. Though some, including Commissioner Curt Hébert, hoped this order would be the final word, FERC signaled that more work needed to be done to reach market consensus on several issues, including a new market structure for the state. With a couple of important exceptions, the Commission essentially gave a rubber stamp of approval to its proposed order that was issued on November 1, after receiving input from participants throughout the energy industry.

As expected, FERC adopted its "soft" rate cap of $150/MWh. In real terms, this meant that bids above this amount could not set the market clearing price that is paid to other bidders for power on the wholesale market.

FERC also officially adopted the policy that removed restrictions against bilateral contracts in California and de-emphasizes spot-market activity. Moving forward, buyers and sellers in the California market would no longer be forced to sell into and buy from the California PX. Instead, long-term contracts may be established between power distribution utilities and power sellers. Specifically, FERC set into place a $74/MWh "target price" for forward contracts. This represented a new policy on the part of the Commission that had not previously been outlined in its proposed order, and was the source of a great deal of contention within FERC. Commissioner Linda Breathitt offered that the $74/MWh target price could possibly set the "standard for reasonableness." However, unlike the $150/MWh rate cap for transactions on the open market, prices below the target price of $74/MWh will likely be accepted without the same level of scrutiny from FERC. Higher prices may be subject to review.

In the proposed order, FERC had proposed establishing independent (non-stakeholder) boards for the California ISO and the PX. This turned out to be more easily said than done, and was one of the earlier proposals from which FERC retreated in its final order. Essentially FERC had wanted to replace existing members of the respective boards with new members not associated with the energy industry. However, reportedly Gov. Gray Davis put pressure on FERC to not adopt this policy because the governor himself wanted to be able to choose new members for the boards that would be "more sensitive to the needs of California consumers."

Signaling that FERC's final order would by no means close the book on California's energy problems, SCE responded with "deep disappointment" to the Commission's order. Specifically, SCE claimed that the order would do nothing to protect California consumers from the unjust and unreasonable wholesale electricity prices that FERC already agreed were in existence in the state. SCE previously had urged FERC to adopt cost-based pricing rules, meaning that each power seller would be able to bid into the market at variable operating costs. FERC had taken this cost-based approach with the initial operation of the restructured Pennsylvania-New Jersey-Maryland (PJM) power pool, which appeared to be working rather well in that area. However, FERC did not adopt this approach for California, and SCE contended that since November the soft rate cap of $150/MWh had been in place and rates had continued to skyrocket.

Again, as we saw with the proposed order, dissension ran high within the FERC itself. Commissioner Hébert reportedly felt that the final order was "too timid" and that it did not do enough to solve the problems in California. FERC's then-Chairman James Hoecker appeared to be open to region-wide rate caps, which Hébert strongly opposed and said could never receive approval from the Commission. Hébert's consistent position had been that FERC is over-extending its reach to mitigate prices, something he believes the Commission was ill equipped to do. Hébert, who would soon become the next FERC chair, disliked any kind of rate cap and preferred instead to let market participants manage their own exposure to price risk.

Perhaps the best feature of this final order was the removal of the requirement that buyers must obtain their power through the California PX. The three California IOUs (PG&E Co., SCE, and SDG&E) represent 80% of the buying market in the state. Through establishing bilateral contracts, they would have a new way to negotiate the prices they pay for power, which could ultimately bring some stabilization to the market in the state. Of course, this approach became moot when the California Department of Water Resources became the sole power purchaser for the state.

The other resolution measure over which FERC had jurisdictional control (and mixed feelings) was the issue of price caps for the Western region and California in particular. In March 2001, reports circulated that Curt Hébert, who by that point had become FERC chairman, said that he would block a move by fellow commissioners to impose a temporary price cap on wholesale electricity prices in the western United States. Hébert reportedly said that even if a majority of the three member commission favored price caps, he would prevent a vote from being taken because he did not believe price curbs would increase electricity supplies or reduce demand in the region. "Absent someone proving to me that price caps would bring more supply and less demand in California and the West, my answer is yes," Hébert said, when asked by reporters if he would block a FERC vote on western price caps. A FERC spokesman later downplayed Hébert's remarks. "He did not say he would block any vote," said FERC spokesman Hedley Burrell. "He said he would need to be shown that price caps would bring more supply or less demand to the West."

FERC was clearly divided on the issue of price caps. Obviously, Hébert has expressed strong opposition to any kind of restriction on pricing. However, Commissioner William Massey has said that he will continue to push for price caps to help calm the volatility of the California market. Linda Key Breathitt, the third commissioner presently at FERC, remained undecided on the issue.

In response to ongoing pressure and an energy crisis that showed no signs of abating, FERC fid finally put price controls into place in California in June 2001. Under the federal order, wholesale power prices would be limited around the clock in California and 10 other Western states following a 5-0 vote by FERC. The order, which lasts through September of 2002, limits the prices power generators can charge to utilities under a complex formula based on the costs of the least-efficient producer during any given hour. It expanded on an April 29, 2001, FERC order restraining wholesale prices in California during power emergencies.

FERC clearly had bowed to the pressure from California officials and the ongoing volatility of Western markets and expanded a previously adopted price screen policy that it believed had already been successful in curbing wholesale power prices. While the federal government had maintained its public opposition to any long-term form of price controls,

FERC had apparently found some wiggle room to create a "price mitigation plan" designed primarily to reduce price spikes in California and other Western states. The Commission offered that it had not imposed cost-based price caps in the West, as demanded by California officials, but rather had established a price-mitigation plan, based on market principles, that would be applied to all of the Western spot markets.

There are some key elements of FERC's order that are important to understand. First, the order was a significant expansion of the Commission's previously established policy for Western price caps that was formulated in late April 2001. Previously, FERC applied price controls only in California and only during times of a Stage 3 power alert (signaling that power reserves had fallen to dangerously low levels). Varying from month to month, the price screen previously put into place was based on what it cost to produce power at the least efficient (and therefore most costly) plant running at the time. Under the expanded order, price controls now would also be used during non-emergency periods throughout the entire Western region (11 states). The other states included in the Western Systems Coordinating Council (WSCC) are Washington, Oregon, Montana, Idaho, Wyoming, Utah, Arizona, Nevada, New Mexico, and Colorado.

The expanded order retained the use of a single-price auction and must-offer and marginal-cost bidding requirements when reserves are below 7% in California. What was different is that now the California ISO market clearing price would also serve to constrain prices in all other spot markets in the Western states, and would also be adapted for use even during times when reserves are above 7%. Until September 2002, during non-emergency periods the price for wholesale power in all 11 Western states cannot exceed 85% of the cost of electricity sold during a Stage 1 (or lowest level) power emergency. The new rule set an initial price ceiling of $107.9/MWh for wholesale power sales, which is considerably lower than the average price screen put into place by FERC's original order. Power generators would not be permitted to sell above the mitigated prices in the Western markets.

In addition, all public utilities that own or control generation in California must offer power into the California ISO's spot markets. This rule also applied to non-public utilities selling into the markets run by the California ISO or using FERC-jurisdictional transmission facilities. Other power sellers operating in the WSCC must also offer power into spot markets in the region, but have more flexibility in choosing among the spot markets among the 11 states.

FERC Chairman Curt Hébert offered that the plan relies on "market-oriented principles" that would restrain prices rather than set them by "bureaucratic fiat." Hébert also said that the tying of price structure to the efficiency of production would encourage power generators to invest in new facilities. As the order stopped short of imposing strict price limits based on the cost of an individual generator's production, the Commission argued that the policy did not represent price controls in its strictest definition.

The most surprising element of FERC's expansion of price controls for the West is the length of time that the price caps would be kept in place. The commission appeared to go from one extreme to the other, moving from strong resistance to any form of long-term or permanent price controls to an order that mitigated prices for more than a year (14 months to be exact). Although FERC attempted to find a balance among the polarized positions in the price cap debate, its order has not quelled the ongoing disagreements related to this issue, with the Bush administration, California officials, and power generators weighing in with disagreement.

As noted, the Commission's order attempted to reach a middle ground by striving to give something to all of the various stakeholders. To some extent, FERC succeeded in reaching this objective. In general, Democrats (most clearly represented by California Sen. Dianne Feinstein) had called for price caps based on each generator's cost of production. Republicans generally resisted any form of price controls, preferring instead to let the market run itself. With the Commission's new ruling, Democrats gained a limited form of the price controls that they sought. Feinstein responded that the order was "not perfect" but did represent "a giant step forward." In turn, Republicans said that market forces will still play a lead role in determining electricity prices.

However, it would be inaccurate to say that the Commission's order was met with overwhelming enthusiasm. Houston-based Reliant Energy, which has been singled out by California Gov. Gray Davis as one of the companies that had unjustly profited from the state's energy market, responded that

FERC's expanded price controls were more of a "political response to the California crisis than an acknowledgement of the market realities in California." Further, the company said that the Commission had ignored the basic principles of supply and demand and reiterated its position that any form of price controls would decrease available supply and discourage conservation on the part of Californians. The Commission's order would only serve to further destabilize the California market, Reliant said. To solve the problem, Reliant believed that California needed a long-term plan that would increase the state's generation supply and provide incentives for reduced demand.

Gov. Gray Davis, while generally pleased that FERC had taken action to control prices in California and the West, also raised concern that the expanded order did not address previous power sales that had led to significant debt on the part of the California utilities. In fact, FERC's expanded order did not take effect until June 19, 2001, and did not apply to any power sales that took place prior to that date. The Commission indicated that it would address refunds for past periods in future orders.

Chapter 6
The Re-Regulation of California's Energy Market

The problems that surfaced with California's direct access program unquestionably have had a profound impact on deregulation efforts across the country and quite possibly around the world. As the first state to launch into electric power competition, California has naturally been viewed as something of a litmus test for how successful deregulation will be nationwide and, on the contrary, how struggles with deregulation may negatively impact a particular market. Other states (and foreign countries as well) that have deregulated their energy markets—or are about to do so—have kept a close eye on the problems in California. In an interesting ramification, these problems sparked a national debate over whether returning to or maintaining a regulated energy market is a better option than proceeding with deregulation.

National Re-Regulation Trend

Just as California dove head first into electric power competition as a means to lower its sky-high electricity prices, states that face high costs in their energy markets remain interested in deregulation. Those states that have comparatively low costs may be more inclined to point to the California experience as another reason why deregulation should be avoided. Whichever the response, it is clear that the events known as the California energy crisis sparked something of a national trend, with equal numbers of states terminating their restructuring efforts and others pushing forward toward electric power competition.

However, perhaps even more pertinent is a movement that developed both within California and other states that arguably is a direct result of the problems seen on the West Coast. This movement is founded on a platform that suggests deregulated energy markets should become regulated once again, and those states that haven't adopted a restructuring plan should take a "timeout" and reconsider whether competition truly offers more benefits than headaches.

Not surprisingly, California lawmakers and consumer advocates suggested that the California market should become "re-regulated." Proposals included ideas from imposing more governmental control over energy prices to having utility companies buy back all the power plants they had sold only two years earlier. Even more striking is a coalition of 23 low-cost energy states that petitioned Congress to exempt them from participating in deregulation. The Low-Cost Electricity States Initiative (LCESI) actually has been in existence for over a year, but gained momentum again now that it can use California as an example of all that is wrong with deregulation. Some of these states only want to open their wholesale markets to competition, but not their retail markets. Others want to avoid deregulation altogether. As a lobbying group, the LCESI has tried to convince Congress that any federal restructuring law should be secondary to state plans, and that any mandate regarding state participation in deregulation will be appealed.

Other low-cost states have started to drag their heels regarding deregulation, even if they have not officially joined LCESI. Iowa, for instance, decided that it would not even pursue the issue of deregulation during the 2001 session of its legislature, upon the urging of the state's largest utility, MidAmerican Energy. New Mexico, which had previously approved a start date of January 1, 2001, for electric power competition, delayed that start date until January 1, 2002, to further investigate the impact that competition would have on the state.

In April 2001, the Nevada Assembly voted to pass Assembly Bill 369, which halts the plant sales and stops restructuring in the state. The key provision of Assembly Bill 369 blocked the sale of power plants before July 1, 2007, ensuring that adequate generation would remain in the state to avert the supply/demand imbalance that had plagued neighboring California. Nevada Gov. Kenny Guinn also pledged his intention to block electric restructuring in the state throughout his term as governor, which ends in 2003, as a direct response to the events in the California electricity market.

The Arkansas Legislature passed Act 324, meaning consumers won't be allowed to choose their electricity supplier until October 2003 at the earliest. The act also established criteria to help the Public Service Commission gauge the readiness of wholesale electricity markets—the underpinning of retail competition.

At the other extreme, two states (Illinois and Ohio) moved along with their restructuring plans in 2001 while continuing to closely watch California. Officials in both Illinois and Ohio remained adamant that the California problems would not occur in their states. Ohio pointed to the fact that its rate freeze would not be lifted for five years, which would protect residents from the kind of price spikes had hurt customers in San Diego. Illinois customers also will benefit from a longer rate freeze and the fact that regulators in the state—having watched problems regarding supply materialize in California—allowed for incumbent utilities to divest of their power plants, but did not mandate it. Thus, importing power from the wholesale market won't be as necessary in Illinois as it is in California.

Between the 23 members of LCESI and the approximately 24 states that have adopted some form of a restructuring plan, the United States is almost evenly divided with regard to states that would or would not be participating in deregulation. But, realistically, would states even be allowed to "opt-out" from electric power competition? Perhaps this might work on the wholesale market, but ultimately it does not seem feasible that half of the U.S. states would retreat from retail competition if the other half is participating. When educated properly about the true benefits of competition, electric power customers—

especially industrial and commercial customers—usually push for competition within their states or go to states where they can choose their energy provider. In addition, because of the way the nation's transmission grid is structured, it would be nearly impossible to transport power any great distance if states along the way refuse to grant access to their transmission lines.

Moreover, perhaps it is not entirely wise to use California as a test example of how deregulation will work nationally, as the LCESI has attempted to do. First, California clearly made some mistakes in its own restructuring plan, not the least of which was the combination of encouraging divestiture and restricting new plants from being built, ultimately leaving the state with power supply problems. Other states are wisely learning from these mistakes and developing better restructuring plans. Second, states such as Pennsylvania have adopted longer periods for rate freezes and stranded cost recovery, thus protecting customers from price spikes for a longer period. Third, direct access in California was terminated after being in effect for only three-and-a-half years. To write it off as a "failure" without allowing this huge system to work out its own kinks was seen by many as a mistake in itself.

Governor Davis Returns California to Regulated Model

The return of California to a regulated model began in early 2001 as a direct result of the extreme financial instability of the state IOUs. Further, re-regulation and a heavy reliance on the state entering all parts of the electricity business emerged as the overriding theme of California's plan to deal with its chaotic energy market. A plan to try to fix the energy crisis by making the state a major electricity broker was introduced by the California Legislature in January 2001 and approved by Gov. Gray Davis. After meetings with energy providers, and state and federal officials, Gov. Davis said the state would try to sign contracts with electricity wholesalers to buy power and sell it to utilities.

When Gov. Gray Davis delivered his State of the State address in January 2001, a major theme of his speech was that California should assume greater control over its own energy market. What became clear was that Gov. Gray Davis, along with the California Legislature, intended to take control over at least some portions of the state's energy market. Specifically, Gov. Gray Davis intended to establish a new state entity charged with the authority to build new power plants, place tighter scrutiny on merchant power generators, and overhaul the bidding process of the California PX.

However, it also became clear in early 2001 that Gov. Gray Davis' plan would extend far beyond these fundamental points. While Davis did not use the word "re-regulation," his efforts clearly amounted to just that as he worked to give back to the state government direct control over energy operations, utilities, and energy usage among California residents. Davis believed the state could negotiate better prices for wholesale power than the state's three IOUs. In particular, PG&E Co., and SCE saw their credit ratings downgraded as a result of their heavy debt loads. As such, state officials sought forward contracts around five cents/kWh, although offers from merchant plant generators still hovered around the seven to eight cents/kWh range. While these figures were higher than the average cost only one year earlier (in the three to four cents/kWh range), they were considerably lower than the high prices that drove PG&E Co. and SCE into financial insolvency (reportedly in the 30 cents/kWh range).

Upon the urging of Gov. Gray Davis, in January 2001 the California Assembly passed a bill by a vote of 60 to 5 that allowed the Department of Water Resources, a state agency, to enter into long-term contracts to buy electricity for not more than 5.5 cents/kWh ($55/MWh). This figure was of course far less than existing market prices or forward prices. The electricity would be resold to utilities or directly to consumers at cost. The bill was signed by the California Senate and signed by Gov. Gray Davis. Immediately, the price limitation element of the re-regulation plan raised concerns that the state had unrealistic expectations regarding its ability to secure reasonably priced power on the wholesale market.

The Department of Water Resources, which had been designated as the state's only credit-worthy buyer, implemented the governor's emergency order of January 17, 2001, and began to negotiate contracts and arrangements for the sale and purchase of electricity to help the state mitigate effects of the electrical shortage. The Department of Water Resources claimed that it was experienced in this task, having done it for many years for the State Water Project. The California ISO, which retained the lead role in distributing energy and making the call on outages, had staff co-located with the Department of Water Resources. The two agencies worked closely under the emergency order.

Subsequently, the Department of Water Resources took over the exclusive role of buying power from generators that would be distributed to California customers by the state's IOUs. The state initiated and signed long-term contracts with a number of power generators that locked in power sales by the state, at times for 10 years. This was a major development and departure from the original tenets of AB 1890, as it put the State of California into the role of buying power. In an effort to accommodate the financial instability of the IOUs, the State of California had transferred enormous financial risk from the utilities to the state government.

In an interesting development that signals an ongoing jurisdictional dispute between the California state government and FERC, the federal commission in late November 2001 ordered the California ISO not to give preferential treatment to the Department of Water Resources over out-of-state generators. Further, FERC also said at the same time that it had jurisdiction over the Department of Water Resources. The new orders from FERC brought on complaints from out-of-state generators such as Mirant and Reliant, which argued that the California ISO had been sharing confidential information with the Department of Water Resources, giving the state department a competitive edge over the power suppliers. The California ISO countered that the Department of Water Resources should be able to obtain non-public information such as the amount of power that an out-of-state supplier had to offer on the wholesale market, because it was buying power on behalf of the three utilities and served as guarantor of third-party transactions for the utilities. This is an argument that FERC has now rejected, and the federal commission has made it clear that the Department of Water Resources should be treated in the same manner as other market participants are treated.

The Pursuit of IOU Transmission Assets

Another key aspect to the re-regulation of California surrounded the state government's interest in purchasing three-quarters of the state's entire transmission infrastructure, which were essentially the transmission assets of the three IOUs. The assets include 30,000 miles of high-voltage transmission lines, countless conductors and insulators, and a string of substations where voltage is altered. The other one-fourth of the grid would remain as it is—owned mostly by government bodies, from the federal Energy Department to the Los Angeles Department of Water & Power.

Statewide, California's transmission grid includes connections to hydroelectric and nuclear plants. Other arms of the system jut across state lines, to import power from Utah, Arizona, or the Pacific Northwest.

Although the idea was not originally part of the legislative motion that gave the Department of Water Resources the right to purchase power, a separate measure on this issue started to float in the California state government, and soon gained the support of Gov. Gray Davis. California's State Treasurer Philip Angelides first proposed that the state take over the electricity transmission lines of the state's IOUs. Under Angelides' plan, $10 billion in state bonds would be sold to buy the transmission systems in the state, as well as finance new power plants. Together, California's three IOUs own about 75% of the state's transmission, although Angelides also proposed that the state assume control of transmission lines owned by municipal utilities such as the Los Angeles Department of Water and Power.

Gov. Gray Davis soon embraced the concept of the state stepping into the transmission side of California's energy business. In late February 2001, Davis announced that he had established an "agreement in principle" with SCE under which the state would buy the utility's transmission lines for an estimated $2.76 billion. The governor said that under the deal he also expected SCE to sell electricity to the state at well below current market rates for 10 years. In return, the utility's parent company, Edison International, would contribute $420 million to the utility to help it defray the billions of dollars in power costs run up in the spot wholesale market. SCE would also have to drop a lawsuit against state regulators seeking rate increases.

For many, the confirmation that the State of California had now officially moved into the transmission business, in addition to its previously adopted role of power purchaser, signed the death certificate for deregulation in the state. In a growing approach that became increasing controversial among other California politicians, Gov. Gray Davis' goal was to gain more control over California's electricity supply, accelerate transmission upgrades, and restore the state's largest utilities to financial solvency. From SCE's perspective, this agreement offered a fairly beneficial resolution to its problems, considering that there appeared to be no other alternative to stave off impending bankruptcy proceedings.

The estimated $2.76 billion price that the state offered to pay for SCE's lines seemed overly high to most observers, and in fact was about 2.3 times the system's book value of $1.2 billion. This had already caused many analysts to speculate that the state would have to receive public subsidies in the near future to support this purchase. However, what drove the proposed purchase was the governor's goal to restore the financial solvency of the state's utility companies without requesting further rate increases or otherwise impacting the state's economy. Under this agreement, in exchange for the money received for its transmission lines, SCE would agree to drop any further legal action it was planning to take to secure additional rate increases. That should have allowed the governor to further reassure California residents that no further rate increases would be necessary to extricate SCE from its staggering debt load. In fact, Davis had said that he did not expect electricity rates, at least in SCE's territory, to increase beyond their existing level (including the temporary increase implemented in January 2001).

In addition, under its agreement with Gov. Gray Davis, Edison International would transfer approximately $420 million, the equivalent of an expected income-tax refund, to SCE, to alleviate the utility's debt and allow it to recover a substantial portion of its uncollected costs associated with spot market purchases. Consumer groups had lobbied to the governor that Edison International should be forced to share some of its subsidiary's financial burden. John Bryson, Edison International's CEO, went on record to reiterate that the utility believed it was entitled to be reimbursed for all uncollected costs and that the deal offered by the governor "is far preferable to perhaps years of protracted litigation."

In addition, Gov. Gray Davis believed that the state would benefit by securing a commitment from SCE to sell electricity to the state at well below current market rates for 10 years. This apparently was a key part of the deal from the state's perspective, as Gov. Gray Davis was able to lock SCE into rates that were based on the cost of production rather than market prices. Although SCE divested much of its power assets under agreements with the CPUC, it continued to share ownership of the San Onofre Nuclear Plant Units 1, 2, and 3 with SDG&E. In addition, a sister company, Edison Mission Energy, reportedly would also provide electricity to the state at "cost-plus" rates from the Sunrise Mission power project, a gas-fired power plant. Reportedly, the Sunrise plant was expected to provide about 320 MW of electricity during peak periods of demand in the summer of 2001, which Davis said could save ratepayers $500 million over the next two years. Along with other agreements that the state had made with power providers such as Calpine Corp. and Mirant (formerly Southern Energy), Gov. Gray Davis could make reassurances to Californians that power supply to the state was being increased, which hopefully would help to avert the extensive power emergencies that had been common.

Gov. Gray Davis said that he was close to reaching similar purchase agreements of transmission assets with PG&E Co. and SDG&E. However, PG&E Co. remained adamant that it would not entertain any offer from the governor to sell its transmission assets. Gov. Gray Davis maintained that he would not enter into any deal to buy transmission lines from PG&E and SDG&E unless they also agreed to drop any pending lawsuits against the CPUC.

In order for the state to gain total control over California's energy system, Gov. Gray Davis realized that he needed to secure deals with all three utilities. Together, the three utilities own and operate somewhere between 26,000 and 32,000 miles of power lines (varying estimates). Collectively, the transmission lines for the three utilities carry a book value of $3 billion. The total cost of acquiring the transmission lines from all three utilities could range from $4.5 billion to $7 billion. However, of the three utilities, PG&E Co. reportedly was the most reluctant to enter a deal in which it would sell its transmission assets. Most likely PG&E was holding out for an even higher multiple of book value of its transmission lines than SCE received, which would make sense as PG&E carried a higher debt load than SCE.

SCE Agrees to the Transmission Deal

In April 2001, at a joint press conference with Gov. Gray Davis, Edison International Chairman, President, and CEO John Bryson announced an agreement plan to restore SCE to financial health. "The negotiated resolution with the governor is far preferable for our company, our employees, and our customers than is going into bankruptcy," Bryson said. The key part of the agreement is that the state will receive a primary utility asset—SCE's 12,000 mile transmission system. SCE employees would operate and maintain the system through a contractual arrangement with the state.

Only days after PG&E Co., its California utility counterpart, declared bankruptcy, SCE announced what was billed as a pact with the State of California that theoretically would keep it out of bankruptcy court. From all appearances, it seemed to be a good deal for SCE, as it would not only keep the utility financially solvent, but would also remove it from a line of business that remained rather uncertain and financially unrewarding.

Edison International agreed to sell SCE's transmission assets to the State of California for $2.76 billion. As noted, the confirmed price tag that the state agreed to pay for SCE's lines seemed like a generous offer because it was about 2.3 times the system's current book value of $1.2 billion (the original cost, less [any accumulated] depreciation (OCLD) recorded in SCE's books). Since transmission rates are based on an allowed rate of return on OCLD, it was uncertain that SCE's current transmission rates would provide sufficient revenues for the state to cover its purchase cost.

The proposed sale included only SCE's transmission assets. It appeared that SCE would remain in the distribution business and continue delivering power to customers and running generation. Both the state and SCE agreed to commit to no less than $3 billion of capital investment in utility infrastructure over the next five years, which presumably included upgrades to the transmission system and new generation capacity. It was not entirely clear how the $3 billion in capital investment would be shared between the State of California and SCE.

SCE Chairman and CEO Stephen Frank expressed confidence that a final agreement could be reached "before the end of the year [2001]," although the California Legislature reportedly had agreed to expedite its own review of the agreement and provide approval before summer temperatures began to wreak havoc on the California market. For its part, SCE said that it was "racing toward" an agreement with the governor to avoid being forced into an involuntary bankruptcy.

An argument could be made that the agreement SCE had reached with the State of the California could have provided benefits to the utility that went beyond protecting it from bankruptcy. One could argue that SCE would have benefited by being able to exit the transmission business altogether, which in many ways could cause tremendous problems for the companies that choose to remain in this business. Transmission operations also typically offer a low rate of return; in SCE's case, according to its most recent 10K filing, its transmission business resulted in a return on equity of 9.68%. SCE had proposed that the return on equity for its transmission assets be set at 11.6%, which FERC rejected (FERC has ultimate authority over transmission service pricing). Thus, the question could be raised of why SCE would even want to remain in the transmission business, given this low return on equity. Behind the scenes, SCE could very well have previously made the determination that its transmission business was not a valuable part of its overall operation. This could be representative of an international trend, as regulatory agencies in Australia and Great Britain also have set lower limits on the average transmission network charges that utilities can implement.

SCE Transmission Deal Pronounced Dead on Arrival

The California state legislature adjourned for the year on September 15, 2001, without approving a last-minute bailout for SCE, prompting Gov. Gray Davis to vow to call the lawmakers back in two weeks for a special session to keep the utility out of bankruptcy. The state Assembly had approved a $2.9-billion rescue plan help the utility recoup some of the costs it ran up in California's energy crisis, but the bill stalled in the State Senate over concern it was too sweet a deal for the utility. "Unfortunately, the Senate has not gotten the job done," Gov. Gray Davis said after the legislature adjourned. "I will call a third special session, which will begin in approximately two weeks, so Edison can avoid bankruptcy."

Concerns of a perceived bailout for the company, along with disagreements about the extent to which the State of California should become involved in the energy business, continued to thwart any agreement among the state's officials. As noted, the California Assembly had previously passed a rescue plan for Edison International that included a five-year option for the state to purchase the transmission assets of SCE for about $2.4 billion, or twice the assets' book value of $1.2 billion. This option became a sticking point as many California lawmakers questioned why the state would want to take on the responsibility of operating the utility's 12,000-mile transmission assets, especially if the assets were overvalued. Also included in the Assembly's plan was a provision to issue bonds in the amount of $2.9 billion to pay off some of SCE's debts (mostly payments to lenders and small power producers, but not to large generators, which the state had accused of price gouging).

The disagreements among California lawmakers were represented further when the state's Senate made significant changes to the Assembly's bill. Among these changes were a lowering of the purchase price of SCE's transmission assets to book value, and lowering the bond allowance to $2.5 billion. Gov. Gray Davis and executives at Edison International immediately responded that the revised rescue plan would be insufficient to stave off bankruptcy proceedings for SCE. In fact, Gov. Gray Davis reportedly said that the revised Senate bill "would not get the job done" and that it was pointless to pass any bill that would not accomplish the objective of restoring Edison International to creditworthy status. Thus, the bill hit a brick wall with the California Assembly not even bothering to debate it, knowing full well that it would not have the required votes for passage in the Senate or be given the required sign-off approval from Gov. Gray Davis.

The challenge for all of the legislators involved would be to see if a rescue plan could be formed that not only was deemed sufficient to keep Edison International financially solvent, but also earn the support of the

California Assembly and Senate. Remember that it was Gov. Gray Davis who established a Memorandum of Understanding with SCE in April 2001, outlining the various steps that would be taken to keep the utility and its parent out of bankruptcy. The deadline for that particular agreement had expired, but Davis remained among the most diligent of California officials in his attempts to keep the "number two" private utility in the state (after PG&E Co.) financially solvent. Of course, PG&E Co. by that point had already declared bankruptcy in April 2001 and was operating under Chapter 11 court protection while it formulated a reorganization plan. Ironically, the ability of PG&E Co. to continue providing uninterrupted electric service while navigating through bankruptcy proceedings caused California legislators to be even more reticent to agree to a rescue plan for Edison.

Ultimately, it was the California Legislature and CPUC that had the authority to approve the agreement between Gov. Gray Davis and SCE. These two agencies were charged with deciding if the plan was good or bad, and weighing the impact for various stakeholders. The legislature in particular found that there were many questionable aspects about the plan. For instance, what value would there be in the California government owning only one portion of the state's interconnected transmission system? PG&E Corp. walked away from the negotiating table and expressed little interest in selling its transmission lines to the state. No agreement with Sempra Energy (parent of SDG&E) had been reached either. Thus, if the Gov. Gray Davis/SCE agreement was approved, the State of California would have gained ownership rights to only a portion of the overall transmission system in the state. Many legislators challenged the value of this, especially when it was considered that most of the congestion that led to higher prices within the state was concentrated along PG&E Co.'s portion of the grid. The counter-argument to this point was that, now that its utility was in bankruptcy proceedings, PG&E Corp. might not have the choice of whether or not to sell its transmission assets to the state. The bankruptcy judge would make that decision. Thus, some argued, the state may have had a better chance of obtaining PG&E Co. transmission assets if it could first complete a deal with SCE.

The second fundamental problem that caused California legislators to turn their back on the governor's plan was what some have referred to as an "obscene bailout" for SCE and its parent company. Some reportedly believed that Gov. Gray Davis rushed into an overly generous offer for SCE out of anxiety caused by the breakdown in his communications with PG&E Co. As noted, the $2.76 billion offered for SCE's assets was about 2.3 times their book value, so many legislators had difficulty reconciling what appeared to be an inappropriate price tag (and, once again, a cost that taxpayers and ratepayers would ultimately have to absorb). Under the existing structure, SCE customers paid a FERC-approved transmission rate based upon a cost-plus formula. If the state paid 2.3 times book value, it would have equated to ratepayers paying two times over for those same transmission lines. In addition, some legislators, including former Democratic allies of Gov. Gray Davis, suggested that SCE should be forced to enter bankruptcy as PG&E Co. had done. Naturally, legislators approached the issue with their constituents in mind, knowing that they would be accountable for how they voted on this plan.

The energy crisis in California had become so politicized that legislators were very reticent to approve any measure that resembled a financial bailout for one of the utilities, especially without obtaining something of value for the state in return. It was Gov. Gray Davis' personal agenda to obtain control over the state's transmission assets, but this was not a priority for other California lawmakers. Thus, the deal was a tough sell without some component that put owning SCE's transmission assets in the state's best interest.

Problems with the Re-Regulation Plan

As discussed in chapter 5, a rescue plan that helped SCE to avert bankruptcy was reached in September 2001. Yet, there are several problems with the re-regulation approach that was pursued by Gov. Gray Davis and the California Legislature that remain as of this writing.

First, although efforts were being made to increase California's in-state power supply, and Davis' plan included expedited measures to bring such generation online, the state still faced a severe power shortage. Regularly in the first half of 2001, the State of California issued emergency power warnings, indicating that power reserves had fallen below 1.5%. Reportedly, California needs several thousand megawatts of new capacity over the next five years in order to alleviate its power supply concerns. At this point, private generators have only 2,000 MW under construction and 4,000 MW that have been approved in addition to that. Consequently, California should remain a net importer of power for at least the next several years. In this ongoing high demand/low supply scenario, the market power currently held by power suppliers in California should remain intact. Moreover, in the near term, there is simply no guarantee that the state government, as a buyer of power, would have any better luck in securing lower prices than the IOUs.

In addition, if the state demands lower wholesale prices, power generators could very well take their business to other states. For instance, Duke Energy, an owner of power plants in California, has stated that it will not build additional plants there if electricity prices are driven too low. By the same logic, power suppliers that do not own plants in California could choose to sell their power in other regions where they would make more money.

This leads to a second concern with California's re-regulation plan. Under the current market structure, PG&E Co. and SCE are financially unstable due to uncollected fees related to the high cost of wholesale power that they purchased to serve their customers. Although the current cycle of high wholesale prices has contributed to severe problems in the state, this cycle will not last forever. For instance, the ultra-high price of natural gas that is currently in place should drop off in a year or two, as a result of new drilling efforts. Thus, any long-term contract that the State of California established in the early part of 2001 could create unnecessary financial risk for the whole state. Gov. Gray Davis apparently realizes this dilemma, because as of this writing the state has expressed a desire to renegotiate power contracts signed with generators and marketers. For the most part, the companies involved in long-term sales contracts with the state of California have resisted any renegotiation of the contracts.

Moreover, if wholesale prices remain high—or conversely, if California's state government establishes long-term contracts with suppliers that turn out to be higher than market rates—the entire state government could be put at risk for financial insolvency. Without question, this poses a far worse case scenario than the financial issues of PG&E and SCE. Although state control over the energy industry in California may appear to have real advantages at this moment, does the move warrant putting the entire state and its 34 million residents at financial risk?

The third problem with California's plan relates to a question of what will happen to PG&E, SCE, and SDG&E if this re-regulation plan comes to fruition. The California IOUs, having divested most of their own generation assets, are now primarily in the transmission and distribution business. If the state assumes control over their transmission assets, what role would PG&E Co., SCE, and SDG&E continue to play in the California market? It would appear that this plan would strip all three of their fundamental businesses.

Here's another question that could exacerbate this situation: Under a state re-regulation plan, what would happen to the nuclear assets still owned by PG&E and SCE? PG&E Co. owns Diablo Canyon Nuclear Power Plants Units 1 and 2, and SCE shares ownership of the San Onofre Nuclear Plant Units 1, 2, and 3 with SDG&E. Under federal law, original construction permit requestors must demonstrate financial solvency to build and operate nuclear units. Such financial solvency must remain intact throughout the life of the nuclear license, otherwise the Nuclear Regulatory Commission (NRC) can revoke the license. Considering the bankruptcy proceedings of PG&E and the general financial status of SCE, the NRC may be left with no choice but to revoke the licenses for both nuclear plants. The question is, who would then take over operational control of the two plants? As this falls under federal jurisdiction, it is unlikely that the NRC would transfer the licenses of the two plants to California's state government or a private agency. The most likely scenario is that the federal government would have to take some form of financial liability for the plants.

Moreover, the State of California would not bring any expertise to the role of transmission owner. From the state's perspective, it would get an asset in exchange for a bailout, which makes its investment in SCE beneficial. However, the transmission business would present unique challenges for the state, such as raising capital for transmission improvements and siting procedures. The state may quickly learn why the transmission business is perceived as being so challenging, and why it is difficult to entice companies to build new transmission lines (especially considering the low rate of return that the transmission business offers). On the other hand, by assuming the role of a transmission system owner, the State of California will be in a position to obtain competitive information about the activities of power generators in the state. This could result in a significant advantage for the State of California if it remains in its current power-buying role.

Chapter 7
The Great Debate Over Price Caps

Philosophical Gulf on Issue of Price Caps

Throughout the twists and turns of the energy crisis in their state, California's leaders called as never before for regional price controls, sparking the biggest debate over energy market intervention since the 1970s. The Bush administration, key federal regulators, and most Western states remained opposed to price caps, arguing controls just make things worse by discouraging needed conservation and power plant construction. But resistance to price controls weakened as pressure grew to staunch a spreading energy crisis.

Despite fairly unqualified resistance from the major energy industry leaders, the issue of price caps for the western United States remained a hot debate within boardrooms and regulatory chambers. Adding fuel to the fire were the incoming financial reports of the major power suppliers that served this area of the country, which indicated that a handful of generators continued to make huge financial profits while utility companies went into the red and customers faced blackouts.

Further, the fact that currently a handful of power generators were able to accumulate unrestricted profits from the market vulnerability in the West struck many as particularly grievous. Financial reports released by the power generators throughout 2001 suggested that a small number of companies benefited greatly from the lack of price controls in this particular region. Earnings reports from companies such as Dynegy, Duke, Enron, Mirant, Reliant, and Williams (to name just a few of the 13 that are most active in the wholesale markets of the West) demonstrated stellar earnings, beating Wall Street estimates and capitalizing on what could only be seen as a seller's market in most of the Western half of the country. Net income for all three companies rose sharply in the first six months of 2001.

Of course, it must be acknowledged that most of these companies continued to reiterate that California sales played a minimal role in their earnings. Rather, they claimed, the upshot was due to cold weather in the Midwest. However, according to company reports, all of these companies mentioned (with the exception of Enron) own a considerable amount of generation in California: Duke 3,351 MW; Dynegy 2,768 MW; Mirant 3,065 MW; and Williams 3,936 MW. Enron does not own assets in California, but its wholesale trading volumes in the Western Systems Coordinating Council (WSCC) region increased by 73% over the course of 2000 to 2001. In addition, a letter to the bankruptcy trustee in the PG&E Co. case revealed that the California utility owes Enron $570 million for back purchases of power, which clearly demonstrated that Enron's involvement in the state was not inconsequential.

Moreover, the dichotomy among the stakeholders in the Western markets could not be more extreme. Utilities operating as providers of last resort, which purchased power on the wholesale market, became financially weakened (some heavily in debt), while a small group of power suppliers wielded what some labeled as market power in the region grew increasingly rich.

So, are price caps the answer to the Western power mess? That certainly became the multi-million-dollar question. As noted, the different camps became quite polarized on the issue. On the side favoring price caps stood the governors of California, Washington, and Oregon, who represent the largest populations within the WSCC. Joining these West Coast leaders were the California utilities (PG&E Co. and SCE). In addition, California leaders from the CPUC and State Legislature also expressed support for some form of regional price controls. The general consensus was that a temporary wholesale price cap should be put into place for a period of two years. The rationale behind the pro-price cap platform was that price controls would protect both customers and the utilities that must purchase power (or, in California's case, the Department of Water Resources), while still allowing generators to make a reasonable (but not excessive) profit. The price caps would be kept in place for only a short period until the supply/demand imbalance in the region could be resolved.

Standing resolutely against any form of price controls was the Bush administration, including Energy Secretary Spencer Abraham. In addition, it appeared that the governors of the nine other states in the WSCC all opposed any form of price caps, as indicated by a signed letter sent to FERC. The argument from the opposing camp was that wholesale price caps would only serve to lower supply by providing less incentives for power suppliers to sell into the state. In addition, those opposed to price caps said price fluctuation provided market signals to customers to lower their demand, which eventually should bring prices back down.

The true wild card in the price cap debate was FERC, which many observers believed could have the final word on the issue. Former Chairman Curt Hébert traditionally voiced strong opposition to any form of price controls. In fact, throughout his tenure as FERC chairman, Hébert said that he would block a move by fellow commissioners to impose a temporary price cap on wholesale electricity prices in the Western United States. Hébert apparently said that even if a majority of the FERC commissioners favored price caps, he would prevent a vote from being taken because he did not believe price curbs would increase electricity supplies or reduce demand in the region. Part of Hébert's concern was that, as he has said, "Washington doesn't do temporary very well," meaning that a price cap put into place by the Commission, even under the guise of being temporary, would probably remain in place for some time.

Moreover, in additional to political differences, there was a basic philosophical difference that divided stakeholders over price caps. Some believed that market forces should be the sole driver for prices and that, eventually, prices would become more reasonable as the market provides solid alternatives. Others said that the federal government had an obligation to step in and bring balance to the market.

One of the key dilemmas over price caps in the Western markets stemmed from jurisdictional uncertainty. Western states looked to FERC and the Bush administration to solve the regional energy crisis, while federal administrators backed away and said that the crisis was a state issue. The question of who ultimately should take responsibility for finding a comprehensive solution to the Western energy crisis has yet to be answered. As state and regional measures up to this point have proved rather ineffective, once again the ball lands in FERC's court to intervene and provide some level of regulatory direction.

Dialogue between Gov. Gray Davis and President Bush illustrated the huge philosophical gulf over price caps. Gov. Gray Davis repeatedly argued that Californians had been unfairly charged billions in unfair and unjust power charges. Thus, Davis clearly expressed his desire for the Bush administration to enforce federal law by making FERC put wholesale price caps into place for a period of six or seven months (long enough to get past the summer of 2001 power crunch). In an interesting development, Davis gained some influential support regarding his plea for price caps from a consortium of 10 economists, including Cornell professor emeritus Alfred Kahn, generally considered the "father of deregulation." In a letter sent to the President, Davis and the group of economists claimed that FERC's failure to act soon on the issue of price caps "will have dire consequences for the state of California and will set back, potentially fatally, the diffusion of competitive electricity markets across the country."

In contrast, President Bush, along with Vice President Cheney, maintained that price caps would not solve the power shortage, and in fact would only do further damage to California's power problems as price restrictions would discourage further production. In addition, Bush argued that the core of California's problems was a supply/demand imbalance, something that price caps would do nothing to solve. Vice President Cheney said that price caps might worsen the situation by limiting supply, resulting in more blackouts for California. Further, Bush argued that California officials (including Davis' predecessor, Pete Wilson) were responsible for the state's power woes and that the only real solution was a combination of increased supply and conservation efforts.

The subtext of the Bush administration's approach to the California power crunch was that the problem could be solved in time, but not with temporary band-aids. Under his federal energy plan released in April 2001, President Bush outlined a plan to increase power supply over the next 20 years, fueled by traditional and non-traditional sources. Bush's approach to California was that, in time, increased production and transmission capacity upgrades would rectify the state's current supply/demand imbalance. Affixing temporary price caps, according to Bush, would do no good and could cause a great deal of damage.

Of course, politics played a key role in the California energy saga. Both Bush and Davis saw their approval ratings dip as a result of the power crisis. A Field Poll released in the spring of 2001 indicated that President Bush's approval rating in California was at 42% (40% disapproving), 12 points behind the national average. Gov. Gray Davis' approval rating in the state was also at 42%, with disapproval coming in at 49% (a sharp decline from more favorable ratings held in January 2001).

FERC Forced to Take Action

By December 2000, the California energy crisis was in full swing. However, the jurisdictional uncertainty over which governmental agency (state or federal) had the ultimate authority to implement resolutions continued, and thus there was no end to the crisis in sight at that time. Ultimately, it became clear that FERC would need to take the lead on measures to repair California's unstable energy market, and as such the Commission bowed to pressure from state officials to implement price caps for the sale of wholesale power.

In mid-December 2000, the Commission unanimously approved a comprehensive order on ways to fix California's troubled marketplace. Though some, including Curt Hébert, who at that time had not yet been appointed as chairman, hoped this order would be the final word, FERC signaled that more work had to be done to reach market consensus on several issues, including a new market structure for the state. FERC previously had determined that the dysfunctional market structure in California had caused "unjust and unreasonable rates" for utilities that must purchase power on the wholesale market, and end-use customers such as those in San Diego who no longer were protected by a rate freeze.

In its unanimous vote, FERC adopted its "soft" rate cap of $150/MWh. In real terms, this meant that bids above this amount could set the market clearing price that was paid to other bidders for power on the wholesale market. The rate cap was referred to as "soft" because sellers could bid above this level and receive their bid if they were dispatched, but anything higher than $150/MWh would not set the price that all other generators would receive. Any generator setting a bid above $150/MWh was required to report their bid to the Commission, and presumably fall under intense scrutiny.

At that same time, FERC also officially adopted a policy that removed restrictions against bilateral contracts in California and de-emphasized spot-market activity. From that point forward, buyers and sellers in the California market would no longer have to sell into and buy from the California PX, ending a key element of the original AB 1890 restructuring law. Instead, long-term contracts could be established between power distribution utilities and power sellers. Specifically, FERC set into place a $74/MWh "target price" for forward contracts. This represented a new policy on the part of the Commission that had not been outlined in previous orders, and was the source of a great deal of contention within FERC. Commissioner Linda Breathitt offered that the $74/MWh target price could possibly set the "standard for reasonableness." However, unlike the $150/MWh rate cap for transactions on the open market, prices below the target price of $74/MWh would likely be accepted without the same level of scrutiny from FERC. Higher prices would be subject to review.

Signaling that FERC's final order would by no means close the book on the problems in California, SCE immediately responded with "deep disappointment" to the Commission's order. Specifically, SCE claimed that the order would do nothing to protect California consumers from the unjust and unreasonable wholesale electricity prices that FERC already agreed were in existence in the state. SCE previously had urged FERC to adopt cost-based pricing rules, meaning that each power seller would be able to bid into the market at variable operating costs. FERC had taken this cost-based approach with the initial operation of the restructured Pennsylvania-New Jersey-Maryland (PJM) pool, which was working rather well in that area. However, FERC did not adopt this approach for California, and SCE contended that since November 2000 the soft rate cap of $150/MWh had been in place and rates had continued to skyrocket.

Dissension ran high within the FERC itself. Commissioner Hébert reportedly felt that the final order was "too timid" and that it did not do enough to solve the problems in California. James Hoecker, who in December 2000 was still FERC chairman, appeared to be open to region-wide rate caps, which Hébert strongly opposed and said could never receive approval from the Commission. Hébert's consistent position was that FERC was over-extending its reach to mitigate prices, something he believed the Commission was ill-equipped to do. Hébert, who by that point way was rumored to become the next FERC chair, clearly disliked any kind of rate cap and preferred instead to let market participants manage their own exposure to price risk.

The lack of consensus regarding how to fix California's dysfunctional market—not only within the Commission, but throughout the industry as a whole—only seemed to be growing. FERC set out with the clear goal of bringing down energy prices for consumers, but it was still questionable if this final order could accomplish that objective. As SCE noted, the soft price cap had done nothing to keep prices down in the state, and concerns about power supply in California continued to increase. In fact, nearly every day during the month of December 2000 the California ISO had issued a Stage One or Two emergency warning, indicating that operating reserves of power in the state were falling to alarming lows.

Price Caps Expanded for Entire West

In April 2001, FERC expanded its price mitigation strategy from California to other Western states and all transactions. Further, the expanded price caps were extended beyond their former application only in times of possible electricity shortages to a 24-hour-a-day/seven-days-a-week timetable in 11 Western states.

In March 2001, FERC began to establish price screens on a month-to-month basis that set the limit for the amount that could be charged for power in California in times of power shortages. In other words, FERC's discretionary price cap, to be applied only when a Stage Three power alert was issued, limited the price at which power suppliers could sell power into the state without regulatory oversight from the Commission. The price screen varied from month to month, but was based on what it cost to produce power at the least efficient (and therefore most costly) plant running at the time.

There were few supporters of FERC's approach. Critics argued that the price cap was only triggered when power reserves were already low, and the limit was linked to the previous day's prices (based on the least efficient plant), which too often allowed most generators to profit excessively. In addition, generators were still allowed to sell power at a price higher than the price screen, as long as they justified their prices with FERC (making this a "soft" price cap that was not regularly enforced). In turn, power generators also disliked the limit, which they believed was too strict.

Nevertheless, FERC, in the words of then-Chairman Curt Hébert, believed that the Commission's "emergency only" price caps had succeeded in bringing prices down in California. Hébert pointed to the fact that electricity prices on the spot market in California fell in April 2001 to below $100/MWh for the first time in 2001 (down from an average of about $300/MWh earlier the year), while natural gas prices had also fallen. However, the true cause for the sudden drop in Western prices was not fully determined, and could just as easily have been the result from a combination of conservation efforts, weather patterns, and increased competition. Ironically, from a broader perspective, this supported Hébert's general philosophy in that, if left alone, high prices ultimately would lead to greater conservation and increased levels of competition among alternative power sources.

The reason for FERC expanding its price mitigation strategy from California only to the entire Western region was simply one of political pressure. For much of 2000, the debate over price controls seemed to be a partisan issue, with Democrats generally favoring price caps and Republicans, most strongly articulated by the Bush administration, generally opposing them. The tables turned in the early months of 2001 due to two factors. First, Democrats gained control in the U.S. Senate, which pushed the issue of price caps to the forefront. For example, Senate Majority Leader Tom Daschle, in an interview with CBS News, said that Congress should "force FERC to do its job" and questioned why the Commission was needed at all if it "doesn't do its job in a crisis like this." In the same interview, Daschle expressed his support for Western price caps without any qualification. The other dynamic that changed was that more Republicans began to put pressure on FERC to act on Western electricity prices, out of fear about potential fallout in the 2002 elections if electricity prices remained high. It was reported that 15 GOP lawmakers sent a letter to FERC, urging the Commission to expand the limited price controls that were in place in California.

Moreover, the price cap issue remained heavily politicized. The positions of some of the players were still crystal clear. President Bush and Vice President Cheney continued to maintain that the administration was opposed "to any type of price-control legislation." The oft-repeated justification for this position was the administration's belief that price controls would artificially reduce the much-needed supply in California by deterring power producers from doing business in the state. Gov. Gray Davis and the California utilities, which in the early part of 2001 were still purchasing expensive power on the wholesale market, continued to plead for tighter price controls.

Given the ongoing dissension among the participants in this debate, once again the focus returned to FERC, which was considered the agency that must make a final decision on the price cap issue. In mid-June 2001, FERC ruled that wholesale power prices would be limited around the clock in California and 10 other Western states following a 5-0 vote. The order, which lasts through September 2002, limits the prices power generators can charge to utilities under a complex formula based on the costs of the least-efficient producer during any given hour. It expands on the April FERC order restraining wholesale prices in California during power emergencies.

Clearly, FERC had bowed to the pressure from California officials and the ongoing volatility of Western markets and expanded a previously adopted price screen policy that it believed had already been successful in curbing wholesale power prices. While the federal government maintained its public opposition to any long-term form of price controls, FERC apparently found some wiggle room to create a "price mitigation plan" designed primarily to reduce price spikes in California and other Western states. The Commission offered that it had not imposed cost-based price caps in the West, as demanded by California officials, but rather had established a price-mitigation plan, based on market principles, that would be applied to all of the Western spot markets.

There are some key elements of FERC's order that are important to understand. First, the order is a significant expansion of the Commission's previously established policy for Western price caps that was formulated in late April 2001. Previously, FERC applied price controls only in California and only during times of a Stage Three power alert (signaling that power reserves had fallen to dangerously low levels). Varying from month to month, the price screen previously put into place was based on what it cost to produce power at the least efficient (and therefore most costly) plant running at the time. Under the expanded order, price controls now will also be used during non-emergency periods throughout the entire Western region (11 states). The other states included in the Western Systems Coordinating Council (WSCC) are Washington, Oregon, Montana, Idaho, Wyoming, Utah, Arizona, Nevada, New Mexico, and Colorado.

The expanded order retained the use of a single-price auction and must-offer and marginal-cost bidding requirements when reserves are below 7% in California. What was different is that the California ISO market clearing price would also serve to constrain prices in all other spot markets in the Western states, and would also be adapted for use even during times when reserves are above 7%. Until September 2002, during non-emergency periods the price for wholesale power in all 11 Western states cannot exceed 85% of the cost of electricity sold during a Stage 1 (or lowest level) power emergency. The new rule sets an initial price ceiling of $107.9/MWh for wholesale power sales, which is considerably lower than the average price screen put into place by FERC's original order. Power generators will not be permitted to sell above the mitigated prices in the Western markets.

In addition, all public utilities that own or control generation in California must offer power into the California ISO's spot markets. This rule also applies to non-public utilities selling into the markets run by the California ISO or using FERC-jurisdictional transmission facilities. Other power sellers operating in the WSCC must also offer power into spot markets in the region, but have more flexibility in choosing among the spot markets among the 11 states.

Former FERC Chairman Curt Hébert offered that the plan relied on "market-oriented principles" that will restrain prices rather than set them by "bureaucratic fiat." Hébert also said that the tying of price structure to the efficiency of production would encourage power generators to invest in new facilities. As the order stops short of imposing strict price limits based on the cost of an individual generator's production, the Commission has argued that the policy does not represent price controls in its strictest definition.

The most surprising element of FERC's expansion of price controls for the West was the length of time that the price caps would be kept in place. The Commission has appeared to go from one extreme to the other, moving from strong resistance to any form of long-term or permanent price controls to an order that mitigates prices for more than a year (14 months to be exact). Although FERC attempted to find a balance among the polarized positions in the price cap debate, its order has not quelled the ongoing disagreements related to this issue, with the Bush administration, California officials, and power generators weighing in with disagreement.

As noted, the Commission's order attempted to reach a middle ground by striving to give something to all of the various stakeholders. To some extent, FERC succeeded in reaching this objective. In general, Democrats (most clearly represented by California Sen. Dianne Feinstein) had called for price caps based on each generator's cost of production. Republicans generally have resisted any form of price controls, preferring instead to let the market run itself. With the Commission's new ruling, Democrats have gained a limited form of the price controls that they sought. Feinstein responded that the order was "not perfect" but did represent "a giant step forward." In turn, Republicans say that market forces will still play a lead role in determining electricity prices.

However, it would be inaccurate to say that the Commission's order has been met with overwhelming enthusiasm. Houston-based Reliant Energy, which has been singled out by Gov. Davis as one of the companies that has unjustly profited from the state's energy market, responded that FERC's expanded price controls were more of a "political response to the California crisis than an acknowledgement of the market realities in California." Further, the company said that the Commission had ignored the basic principles of supply and demand and reiterated its position that any form of price controls would decrease available supply and discourage conservation on the part of Californians. The Commission's order will only serve to further destabilize the California market, Reliant said. To solve the problem, Reliant believes that California needs a long-term plan that will address increasing the state's generation supply and providing incentives for reduced demand.

The "New" FERC May Become More Interventionist

In August 2001, FERC Chairman Curtis Hébert, submitted his resignation in a letter to President Bush. The resignation took effect August 31. The resignation had been expected amid indications that the president preferred Patrick Wood III, former chairman of the Public Utility Commission of Texas, who assumed the role of FERC chairman in September 2001.

The resignation of Curt Hébert from his post as FERC chairman did not come as a surprise as it had been a part of the industry rumor mill for months. However, it did represent huge news, because Hébert's resignation could be the start of a philosophical sea change at the Commission, given the different regulatory approaches between Hébert and his successor Pat Wood. In a nutshell, the change in leadership could be the start of a new phase for FERC, in which the Commission becomes far more interventionist in wholesale markets and more apt to implement price controls when markets go awry. If this is the case, it would represent a dramatic departure from the regulatory position of Hébert, who primarily espoused a lassez faire philosophy and often reiterated his belief that FERC should be focused on regulating less, not more. Pat Wood, who joined the Commission in May, also appears to embrace free market principles for the energy industry, but seems to be more receptive to asserting FERC's regulatory position when necessary, which of course could have a dramatic impact on how the energy industry continues to become deregulated.

Without question, Hébert's tenure as FERC chairman was marked with controversy. During the last six months of his tenure on the Commission, FERC ascended from being a rather obscure federal agency to a leading voice within the California energy crisis and a target of criticism by other players that claimed FERC had not done enough to stabilize volatility in the West. In addition, Hébert was criticized repeatedly for not taking a harder line against power generators accused of price gouging in California and enforcing refunds back to the state. Price controls, which Hébert had adamantly and consistently opposed throughout his time with FERC, clearly became a lightning-rod issue for both the Commission and other departments within the federal government.

Under the leadership of Pat Wood, many suspect that FERC will become far more interventionist in markets that are becoming deregulated. For instance, Wood has made no bones about the fact that he embraces a market philosophy that includes regulatory intervention when necessary, a position dramatically opposed to Curt Hébert. At a session before members of the Senate Energy Committee, Wood reportedly said that if lawmakers chose to adopt the more traditional approach where regulators set rates based on a producer's cost of service, that he would not be "allergic to that sort of remedy." Consequently, Wood could very well spearhead a re-examination of FERC's methodology regarding market-based rates in markets that become deregulated, along with how merchant plants calculate their rates. This re-examination in and of itself could represent a new era for FERC and a whole new set of standards to which participating companies in the energy industry would have to conform.

In the fall of 2001, the regulatory agenda of new FERC Chairman Pat Wood was beginning to take shape. It is becoming clear that one critical issue for the Commission will be potential market manipulation that might take place by various participants in deregulated environments. One of the ongoing debates related to the failure of California's deregulated market was whether or not out-of-state generators or state-affiliated entities such as the California ISO and the Department of Water Resources intentionally manipulated the state's market to impact power prices. Wood has indicated that FERC will soon release a formal policy for policing companies authorized to sell power at market rates and utilities that use transmission services bundled into state-regulated retail rates to thwart wholesale power competition.

Moreover, the issue of price caps is by no means resolved. At present, the expanded price mitigation policy that FERC initially enacted for California only is now in place for the entire Western region and will remain intact through September 2002. After that point, it seems certain that FERC will once again have to readdress the price cap issue and determine its own level of involvement for bringing market stability to regions such as California through interventionist measures such as the use of price controls.

Chapter 8
Comparison of California to Other States

As noted in chapter 7, some states have chosen to delay or altogether terminate their plans for electric power deregulation, in large part due to the problems associated with the California market. At the other end of the spectrum, two states—Pennsylvania and Texas—have followed through with full competition, continually espousing confidence that their competitive models are inherently different from the one in California and therefore stand a better chance of long-term success.

In fact, the three biggest states to formulate deregulation plans— California, Pennsylvania, and Texas—have all approached competition from dramatically different perspectives. An analysis of the nuances included in the Pennsylvania and Texas restructuring plans reveals the value of retaining the integrated utility model (and the risks of dismantling it).

PENNSYLVANIA

The Commonwealth of Pennsylvania opened its electric power market to full competition only a few months after California, but has had a decidedly more positive experience that the Golden State in implementing direct access. Three years after the onset of electric power choice in Pennsylvania, the state is still viewed by many as offering the most viable competitive market in the United States, if not the world. The wide disparity between Pennsylvania and California was repeatedly highlighted by Pennsylvania officials throughout the California energy crisis. For instance, in January 2001 the chairman of Pennsylvania's Public Utility Commission (PUC) criticized local utility GPU Energy for deliberately alarming consumers and elected officials by suggesting that the energy crisis crippling California could easily affect Pennsylvania. "I am outraged that GPU would even hint that a similar energy crisis could happen to Pennsylvania," said PUC Chairman John M. Quain. "I assure consumers and legislators that what is happening in California will not happen in our state."

From all appearances, it seems as if the Pennsylvania PUC's assurances that the state would not follow California's path into an energy crisis were well founded. However, it is important to remember that Pennsylvania constructed a very different model than California. One of the most important distinctions is that, while the CPUC pressured the three California IOUs to divest generation in an attempt to avoid any market power that the three utilities would have in a deregulated market, Pennsylvania regulators took a different approach. Divestiture was permitted in Pennsylvania, but no special restrictions were placed on how a utility procured power to meet its customers' needs. This approach in Pennsylvania created an altogether different model from the one in California. The Pennsylvania PUC made a special point of ensuring that adequate generation reserves existed to maintain reliable electric power service in the state. Thus, although utility restructuring needed to include separate business units (for example, no sharing of employees), it was really left up to the utilities themselves to achieve separation. In fact, Pennsylvania established a fairly opened-ended plan regarding the retention of the integrated utility model, and left the details to the utilities.

All of the utilities operating in Pennsylvania were required to submit restructuring plans, in which they mapped out their own approaches to competition. Some of the Pennsylvania utilities submitted restructuring plans that included the separation of their generation businesses, yet this was typically done as a strategy to become a registered generation company. According to the Pennsylvania PUC, nearly all of Pennsylvania's electric utilities still own most of their generating plants.

This is one of several factors that makes Pennsylvania a very different model from California. In addition, according to reports released by the Pennsylvania PUC, Pennsylvania produces more power than it consumes, which as a result has helped the state to avoid a supply problem like the one witnessed in California. Pennsylvania remains a net exporter of electricity (California is a net importer), and reportedly is the second largest producer of electricity in the United States. Members of Pennsylvania-New Jersey-Maryland (PJM) Power Pool, the rough equivalent to the California ISO, have a combined generating capacity of 58,000 MW, and increased their capacity by 6,000 MW in the 1990s, keeping up with a 1.4% annual increase in demand for electricity in the Mid-Atlantic states. In 2001, PJM member companies were expected to increase generating capacity by another 3,000 MW. In contrast to the rather restrictive limitations placed on the construction of new power plants in California, Pennsylvania has allowed for new plants to come online more quickly, adding to the generation supply in the state. The result was that, as California faced reserves that dipped below 7%, 5%, and sometimes 1.5%, the PJM power pool had about 35% excess capacity.

Another important factor is that Pennsylvania allowed the state's utilities to enter into long-term contracts with power suppliers. In California, utilities were forced to purchase their power on the spot market (through the California PX), which by definition only offered market prices. As the low supply/high demand equation drove up costs in California, the state's three IOUs had no other option but to purchase power from the California PX at unprecedented high prices.

For certain utilities, Pennsylvania also set a higher default price, or "shopping credit," than the prevailing wholesale price. This has allowed for alternative companies to enter the state and offer competitive prices, which has resulted in more competition. The shopping credit (also known as "price to beat") varies from one utility to another, but on an average basis in Philadelphia for residential customers, it is about 5.65 cents/kWh. If a customer secures a deal under that amount, it results in savings for the customer. If customer locks in a deal higher than that amount, then the customer is paying extra.

Moreover, although California and Pennsylvania opened their electric power markets at roughly the same time, they are vastly different competitive models. In a nutshell, Pennsylvania is not having the same problems regarding supply as California has had, due to a collective decision among stakeholders in that state to allow utilities to retain generation. By having more control over the management of their supply needs, Pennsylvania utilities as a whole should not find themselves in the same predicament as their California counterparts. In addition, longer rate freezes and a higher default price should protect retail customers from the price spikes.

The numbers alone make a strong case for the Pennsylvania model. In figures reported in mid-2000, over two million out of Pennsylvania's five million residential and commercial electric customers have enrolled in the state's customer choice program, giving them the option of electing a new electric service provider. Of that, reportedly over 500,000 customers (concentrated in the service territories of utilities PECO and Duquesne) have actually switched to an alternative supplier. Those figures can be compared to data compiled in California from mid-2000. Two years after the state opened its electric market to competition, barely 100,000 out of California's 10 million electric customers had switched companies. Energy customers in Pennsylvania can look forward to savings as much as 17% off their bills. Consequently, there seems to be a great deal of validity to the claims that Pennsylvania represents the most successful deregulation plan in the United States, if not the world.

TEXAS

Texas officials have been watching California since the launch of competition in the Golden State in April 1998, and to this day remain confident that the power supply problems faced there won't happen in Texas, where full electric power competition began on January 1, 2002. As Texas watches California, the eyes of the energy industry remain on Texas, which is the largest state to proceed with electric power competition in the post California-crisis era. Perhaps most important is the sheer size of the Texas market—which presumably will translate into huge profits for those energy providers that gain a lock on the market—and the hope surrounding how competition in Texas might develop.

After so much negative press about the flawed system in California, a great deal is hinging on how effectively deregulation unfolds in this state. Moreover, the onset of competition in Texas is a major development for the energy industry. With regard to population, Texas represents the second largest energy market (behind California), offering a market value of about $20 billion.

The restructuring law in Texas (Senate Bill 7) seems to reflect a more balanced position when compared to California and Pennsylvania. Indeed, Texas—which was the third of the three states to finalize its restructuring plan—purposefully adopted the elements it liked from both states and discarded what it felt was not favorable.

The Texas law is unique because it requires that each utility separate its regulated utility, generation, and retail energy service activities. Consequently, by January 1, 2002, each electric utility operating in Texas must separate its business activities into the following units: power generation company, retail electric provider (REP), and transmission and distribution (T&D) company.

In addition, Texas law requires that no supplier may own and control over 20% of the generation available to serve a region that is considered "qualified" for competition. As a result, utilities in the state—including TXU, CSW (now joined with AEP), and Reliant—have sold or have decided to sell some of their generation assets. In addition, the power generation company of each unbundled IOU must sell entitlements to at least 15% of its installed generation capacity. In contrast to the PX model established in California, Texas utilities are encouraged to establish bilateral contracts with suppliers to secure the most competitive package. There is no formal exchange or auction process that has been mandated in Texas.

Texas also has its own form of a "price to beat." From January 1, 2002, until January 1, 2007, a REP in Texas must make available to any residential and small commercial customers rates that are 6% less than rates that were in effect as of January 1, 1999. This policy creates a situation in which an REP cannot compete in its own service territory at a price different than the unregulated price to beat. The rationale behind this appears to be that the REPs were allowed to inherit all of their customers, and therefore were not allowed to compete in their own service territories so that competitors could emerge and offer a better a price. The rate freeze will be lifted on January 1, 2007, or before if a REP can prove that it has lost 40% of its customer base.

One important distinction between the two states is that Texas regulators are allowing for power plants to be built at less than half the time it has taken to build a new plant in California. Since 1995, 22 new plants have started operations in Texas, putting out 5,700 MW. By the time full competition started in January 2002, 15 more plants and 10,000 MW are scheduled to come online. As a result, Texas officials say that—unlike California—their state has more than enough generating capacity to meet peak demand and provide at least a 15% reserve for the next few years.

However, there are remaining concerns about Texas' restructuring law. First, a generation company cannot directly sell to customers. In addition, a REP cannot own generation. This eliminates the possibility for the integrated utility model, although a company can financially construct the same capability through the use of bilateral contracts. Will this ultimately contribute to a power supply problem such as the one faced in California? The second concern about the Texas model is the arbitrary 6% rate reduction. How do regulators have any way of projecting what wholesale prices will be?

In addition, although the market potential for Texas competition appears very strong, there are still some lingering concerns about competition in the Lone Star State. First, there is a shortage of transmission lines in Texas, especially in the northern half of the state. Although Public Utility Commission of Texas (PUCT) officials say that sufficient transmission lines will be put into place before competition begins, without sufficient lines, companies could exert monopoly control over electric prices within certain regions.

Another concern is that Texas may be relying too much on natural gas. Almost all of the new plants that have been built in Texas are powered by this fuel source. While natural gas has its advantages (lower emissions than coal, lower construction costs, etc.), presently the price of natural gas is running high. If this trend continues, the anticipated lower cost of natural gas may not be very reliable. Yet, ironically, customers may still benefit because of the multitude of power plants within the state. Abundant supply in Texas may keep the cost of power low for end users, but less profitable for suppliers.

There have also been concerns about the ability of the Electric Reliability Council of Texas (ERCOT)—the rough equivalent to the California ISO—to handle the additional transmission load once competition begins. The fact that ERCOT is building two new facilities, along with expanding its staff, is probably a direct result of these concerns. More than anything else, Texas wants to avoid the horrible press that California has received this summer and to reassure its residents of the benefits of competition.

Nevertheless, former PUCT Chairman Pat Wood (who assumed the role of FERC chairman September 2001) is confident that Texas has done its homework well, and that the state's overall deregulation plan has set the stage for a successful retail competitive market. Further, Wood has an interesting perspective on what complicated the state's energy deregulation market. Basically, Wood says, the California market worked pretty well until it stopped raining in California, Oregon, and Washington. In Wood's opinion, the fact that California is dependent for one-quarter of its power supply on an intermittent source (in this case, hydroelectric power) makes little sense and is largely to blame for the supply/demand imbalance still wreaking havoc in the state. The lack of power plant investment exacerbated the dynamics of the market when a critical piece of the power supply dried up last summer.

Wood's criticism of the California mismanagement focused on the "it's not my job" mentality, considering that three state agencies were involved in California deregulation and yet no specific agency seemed to be in charge. "[George W.] Bush would have run me out of Texas if we had done anything like that," mused Wood. "Our job is to not only do our job today but to plan for the future to look ahead. Granted, there had been no power plant investment, but not looking ahead with accountable facts was an inexcusable critical piece of planning missed by the California Energy Commission."

The perfect storm of colliding conditions that plunged California into survival mode could have been prevented. The whole energy market has a lesson to learn, says Wood. And Texas has an opportunity to practice. To counteract the mistakes made, Wood freely acknowledges that Texas "pirated" most of the infrastructure of Texas' deregulation bill from Pennsylvania's largely successful plan. The Texas legislature meets every two years for a 90-day session. Right before the 1999 session, Wood and a deregulation team went to Pennsylvania and visited with state commissioners, legislators, industry executives, consumer groups, and customers to find out what was working and what was not in the Pennsylvania model.

What they decided to leave behind from Pennsylvania's overall successful deregulation strategies was the provision that allowed a Philadelphia electric affiliate to go back into Philadelphia and attract the customers away from their own utility in the name of competition. Texas' goal, Wood clarifies, is to have a market of unaffiliated companies participating, which will hopefully generate genuine, long-term competition. Yet, in contrast to Pennsylvania, Texas adopted a policy in which customers that do not opt to switch to a new provider are defaulted to an affiliated REP of their incumbent utility. This affiliated REP, which cannot be the T&D operation of the incumbent utility, will remain subject to regulated rates in its assigned service territory, although it is allowed to freely compete without price regulation outside of its service territory.

Chapter 9
Summer 2001:
The Energy Crisis Fizzles Out

A surprising development occurred in California's energy market during the summer of 2001. The energy crisis, which had garnered such intense media scrutiny and wreaked such havoc in the state for over a year, suddenly dissipated. The California ISO ordered utilities to impose rolling blackouts seven times during the first five months of 2001, which certainly heightened concerns about what the subsequent summer would bring. However, despite warnings predicting that the summer of 2001 would bring conditions reaching disastrous proportions, the three months between July and September 2001 came and went without a single electricity blackout. Further, energy prices began to drop considerably, providing a well-needed break to the state of California, which by that time was in the role of power purchaser. End users who feared that existing price freezes would not protect them from escalating rates also got a break.

While the energy crisis in California had by no means ended, and fundamental problems with the state's electricity market still persist as of this writing, it was very clear that the extremely unstable conditions that had become common during the previous summer had begun to subside. This came as a surprise to both state and federal officials, and not only because the general consensus had been that the conditions of the summer of 2001 would worsen. More importantly, California still suffered from an inescapable supply/demand imbalance that should have resulted in the continuation of power emergencies. So what caused the California energy crisis to shift from boiling to a slow simmer during the summer of 2001? The answer lies in many factors that will be discussed in this chapter. However, first it is important to take a step back and recall the dire warnings that caused tremendous concern for Californians in the preceding spring.

The Warnings

In early summer 2001, the California ISO and the CEC acknowledged that demand would exceed supply capacity within California during the subsequent summer months. Specifically, the two organizations agreed that rotating electrical outages would be required, and estimated that the outages during the time period of June 1 through September 30 would range from a low of 55 hours to a high of 700 hours. Further, the two agencies estimated that the involuntary peak demand reduction during electrical outages would range from 1,825 MW to as high as 5,500 MW.

The CEC projected a peak demand for California during June to September 2001 of 47,703 MW. With a desired 7% operating reserve, a total generating capacity of 50,303 MW would be required. For the CEC, the 50,303-MW baseline generating capacity requirement represented a one-in-two chance of occurring, while the one-in-five chance of warmer weather raised the generating capacity requirement to 53,104 MW. What this meant in practical terms was a forecasted shortfall of at least 5,000 MW during the summer of 2001, and maybe as much as 7,000 MW.

Looking at the state from a national perspective, the North American Electric Reliability Council (NERC) issued its own report, the "2001 Summer Assessment," released on May 15, 2001. In the report, NERC made an assessment of the operational conditions and sources of expected problems that could affect the interconnected power grids covering all of Canada and the contiguous United States. Regarding California in particular, NERC indicated that there would be higher customer demand and constrained electrical supply on the Western Power Grid, with rotating electricity blackouts likely to occur in California. For summer 2001 specifically, NERC identified several potential sources for the problems in California.

- Higher summer temperatures, which would increase customer electrical demand;
- Hydroelectric generation limited by lower water levels and reduced snowpack;
- Planned new generating capacity that would not come online when expected;
- Forced outages of generating capacity that would be more extensive than expected; and
- Financial problems, including non-payment to generators, would lower output.

The California ISO and NERC reports started from the same peak load baseline of 47,703 MW and reached the estimated 50,303 MW capacity requirement by adding 2,600 MW of desired operating reserves. However, the California ISO and NERC assessments of generating capacity differed in two important areas:

- The California ISO expected the loss of 2,500 MW due to forced outages. In contrast, NERC looked at a rolling average of the last five years (instead of only summer 2000) and nearly doubled the forced outage rate to 4,525 MW.
- For new generation sources, NERC expected new power plant capacity to grow by 500 MW per month until it reached 1,500 MW in September 2001. The California ISO expected to see new power plant capacity of 3,371 MW by September 2001. The difference between these assumptions was that NERC did not include any new sources of power plant capacity not already sited, under permit, or under construction, and it also considered that the majority of planned capacity would not be available until late in the summer of 2001.

Gov. Gray Davis, feeling continued heat and pressure from both his constituents and national observers to resolve the energy crisis, concurred with these projections and issued his own warnings. Despite the somewhat reassuring announcements that the state of California had lined up various new long-term contracts for the supply of power, in the first half of 2001 it remained questionable whether the power secured in the contracts would be enough to prevent another long, hard summer for the state and its residents. As of the spring of 2001, the state of California was spending upwards of $50 million a day (a total of $2 billion since January) to buy electricity on the spot market. The long-term contracts in which the state had become engaged were intended to lock in cheaper prices and cut down the amount of electricity the state had to buy in the volatile short-term market.

However, speaking at a press conference in Los Angeles on March 5, Gov. Davis announced that state Department of Water Resources negotiators had thus far been able to meet only between 65–70% of the state's power needs for the coming summer, based on consumption patterns from the summer of 2000. The Governor again called on residents to cut usage by 10%. Despite the tempered optimism associated with these announced long-term contracts, the general consensus was that, however much power Gov. Gray Davis had lined up for the summer, it wouldn't be enough. In fact, when pressed for details about how much power would be available that summer, Davis conceded that the state would still have to buy 30–45% of its power on the expensive short-term market.

During the month of August—typically the hottest month in California—the average peak daily demand for power in the California ISO territory is about 47,700 MW. The state's three IOUs—PG&E Co., SCE, and SDG&E Co.—still generated about 8,000 MW from their remaining power plants and had long-term contracts from wind, solar, and other energy sources that gained them an additional 11,700 MW of power. Gov. Gray Davis had established that about 7,000 MW from the long-term contracts would be available during the summer of 2001. In addition, back in February 2001, Davis announced a plan to expedite new power plants in the state, which he promised would bring 5,000 MW online by July 2001. Altogether, assuming that all of these projections were reliable, the total came to about 31,700 MW of power that California should have had available for the summer of 2001, leaving the state about 16,000 MW short for its power supply needs. Requesting assistance from California consumers, Gov. Gray Davis suggested that if Californians would reduce their electricity usage by 10%, conservation efforts across the state could reduce peak demand by nearly 5,000 MW.

The federal government, including President Bush and the Department of Energy Secretary Spencer Abraham, weighed in with warnings about the anticipated events of the summer season. In the spring, the Bush administration warned that electricity blackouts in California "appear inevitable" during the summer months, but continued to issue a strong statement of opposition to addressing the problem with wholesale price caps.

Energy Secretary Spencer Abraham spoke at the U.S. Chamber of Commerce and acknowledged that California clearly would not have enough power supply to meet demand during the summer of 2001. Abraham said that the state would need about 61,000 MW to meet summer demand, while only 56,000 MW of generation was expected to be available. "There's really only three things you can do about a difference between supply and demand," Abraham said. "One is to conserve more. The second is to import more. And the third is to produce more. I think we want to focus more on conserving more and producing more rather than depending more on other countries."

Moreover, regardless of how the various data shook out, most everyone involved in the California energy market agreed that the prognosis for the summer of 2001 was not good. Rather than questioning if additional blackouts would occur, the ongoing debate related to how severe the anticipated outages would become and how long they would last before additional power supply could be established in the state.

The Turnaround

Despite these dire warnings, the summer 2001 fortunately turned out to be a crisis-free season, with none of the major blackouts that had been predicted. Only several months earlier, the state had entered into the summer amidst threats of repeated reliability problems in its power systems.

However, the state essentially was free of power outages all summer long, and the general consensus in the industry was that California's power crisis had significantly subsided. Among the handful of factors cited as the cause of California's newly stable market were conservation efforts put into place by the state's three IOUs. While conservation efforts typically run a cyclical course (increasing when power supplies are compromised or prices spike), the new reports regarding California's efforts also signaled what could be renewed interest for demand side management, voluntary load curtailment, and real-time pricing programs in some areas of the country.

Further, the issue of natural gas prices remains very subjective. Naturally, those in the natural gas production business would like to see prices rise once again, while those on the purchasing side believe the current fall in prices below the $4/MMBtu range is not low enough. Along with this value discrepancy, debate also continues over how long gas prices will remain below $4/MMBtu, with many analysts arguing that the drop is only temporary.

Another result of the drop in natural gas prices is the increase in storage supplies. Traders expect to see a large build according to the American Gas Association. Early predictions suggest a national build of 90 Bcf to as high as 120 Bcf, which is considered bearish for the gas market. Reportedly, year-on-year storage totals for gas supplies already are above a year ago, and gas supply is expected to overshoot demand through 2002.

On the other hand, market forecasters have identified several factors that could cause natural gas prices to rise again. For instance, prices could jump significantly as insufficient pipelines become clogged once again with rising demand (especially in California). Also, weather patterns in the summer of 2001 were comparatively mild. Hot weather in the summer of 2002, 2003, or beyond (and, conversely, cold winters) could cause demand to shoot up once again, causing the fundamental problems of California's energy market to reemerge.

Further, the dramatic increase in new natural gas-fired power plant production that is anticipated across the country (a plan that has been endorsed by the Bush administration) will cause demand to increase substantially, driving up gas prices once again. Most of the 15,000 MW worth of new power plants that are in planning stages will be fueled by natural gas, and consequently should significantly increase energy demand across the country. Demand and high prices for natural gas naturally go hand in hand. With few potential alternatives, natural gas is the fuel of choice for new power plant construction. This can only put upward pressure on gas prices as the demand increases.

In addition, the ongoing drought in the Pacific Northwest will continue to compromise hydroelectric supplies in that region, which could further exacerbate the region's demand for gas-fired generation. Analysts suggesting that California should not get used to the lower gas prices further point to the fact that the state still has an 18%-year-on-year deficit in regional inventories, which keeps the region's natural gas supply in a compromised position.

Despite this debate over price trends for the next 12 months, we know that the first six months of 2001 brought a dramatic drop in prices. The decrease has seemingly resulted from four different factors, which I'll discuss briefly.

The weather has been comparatively mild. Weather is the most important factor that has caused the drop in natural gas prices. Generally speaking, throughout much of the country, spring brought comparatively mild temperatures, which in some places continued into summer. Cooler temperatures meant reduced use of air conditioners in many parts of the country. As natural gas now is being used to fuel a large number of electric power generators, the mild-temperature trend contributed to lower prices.

Demand for natural gas dropped significantly. The second factor is a significant drop in end-user and utility/industrial demand for natural gas. High prices that were common through the early part of 2001 sparked increased conservation levels on the part of end-users and caused many electric utilities that traditionally used natural gas to seek out other fuel sources, such as coal. In fact, electric utility demand for natural gas reportedly has fallen by an average of 8.3% in the first six months of this year. Power consumption was reportedly down about 6.5% in California. In addition, as prices continued to rise, many industrial plants reduced their production levels or even shut down temporarily to avoid paying high power costs. In the Pacific Northwest, for instance, many aluminum companies engaged in contracts with large providers such as the Bonneville Power

Administration to actually shut down production, causing a dramatic decrease in demand. This decreased demand contributed to the drop in natural gas prices, and is actually expected to continue for the near term. The Energy Information Administration (EIA) projects that U.S. demand for natural gas will grow by 1.6% in 2001, as compared to 5% in 2000.

Storage levels increased. A significant increase in storage levels also helped to keep gas prices comparatively low. The combination of mild weather and low prices caused excess levels of natural gas supply, which subsequently was put into storage. According to the EIA, natural gas storage is in surplus compared to last year's level at this time, and as of the end of June is even above the five-year average. Specifically, stored natural gas levels are reportedly about 11% higher than they were last year and 3% above the five-year average.

Production levels increased. The high price of natural gas that was common through much of 2000 sent many production companies scurrying to increase their levels of natural gas output. Ironically, this caused record-high levels of supply, which kept prices low. Demand gas production, according to the EIA, was estimated to have risen by 2.4% in 2000, and is forecast to continue to increase by 3.6% in 2001 and 2.9% in 2002.

Voluntary Load Programs

Ironically, the crisis-free summer of 2001 sparked new concerns for California officials. First, some fear that energy customers will once again become complacent about the amount of power that they use. Some customers have questioned whether or not a power crisis truly exists in the absence of any direct impact on their lives, and consequently the recent conservation efforts may have peaked. Second, state officials are becoming concerned about a power glut in California now that conservation efforts have reduced demand. The state reportedly locked itself into contracts to purchase 43% of the electricity that the three IOUs need for their combined 10 million customers. However, the utilities only need about 35% of the power to meet current demand, which represents an excess of power that must be sold, sometimes at lower cost. According to a report in *The Los Angeles Times*, in June the state of California racked up some $46 million in losses after selling surplus power for one-fifth of the price it had paid. Most of the contracts that the state has signed have 10-year life cycles, and the surplus in power resulting from these contracts is expected to peak in 2004, causing a financial burden on the state unless demand rises substantially once again.

Nevertheless, the conservation efforts in California could signal a re-emergence of demand-side management (DSM) programs that were common in the early 1990s. DSM programs consist of the planning, implementing, and monitoring activities of electric utilities that are designed to encourage customers to modify their level and pattern of electricity usage. Interest in DSM programs decreased dramatically in the mid- to late-1990s as the general consensus in the energy industry believed that deregulation would generate open-market price signals, which consequently would replace the need for DSM alternatives. However, the volatility in some U.S. markets and the resulting high rates following the onset of electric power competition has sparked a renewed interest in DSM programs that provide end users with greater control over their energy usage.

According to information obtained from the Department of Energy in 1999 (most current information available), 848 electric utilities reported having DSM programs. Of these, 459 are classified as large, and 389 are classified as small utilities. This is a decrease of 124 utilities from 1998. DSM costs were almost unchanged at $1.4 billion in both 1998 and 1999. This represented a drop from 1998 figures, in which 972 electric utilities reported having DSM programs. In addition, DSM costs have continued to decrease, from $1.6 billion in 1997 to $1.4 billion in 1998. In 1999 (most recent data available), spending for DSM programs remained flat at $1.4 billion, according to the DOE.

Even prior to the energy crisis of 2000-2001, the three California utilities had developed measures to reduce demand among certain customer groups. SCE, for instance, has had a voluntary load curtailment program in place for many years. Voluntary load curtailment programs are agreements between utilities and certain customers, in which the customers agree to turn off large loads for a specific period of time when the overall grid demand is highest. The utility notifies the customer in advance of an anticipated system peak and offers to pay them to reduce their energy use for a specific period of time. If the customer believes that their inconvenience is worth the offered price, they will accept the offer and reduce their demand for the specified time period. Voluntary load curtailment programs are becoming more common because they are a preferable alternative to interruptible rate service, in which customers are forced to turn off their power service or experience power outages. Voluntary load curtailment appeals to both the customer and the utility because it offers significant advantages to both parties. The utility wins because it can stabilize the grid cheaper than buying additional supply reserves. Customers win because they can sell back their reserved demand at a profit.

The use of voluntary load curtailment programs appears to be increasing as a result of volatility in certain U.S. markets (particularly the West). For instance, in early 2001, when California was regularly experiencing emergency power alerts (indicating that reserves had fallen to a dangerously low level), the California ISO called on the state's IOUs to begin voluntary load curtailment programs for certain customers within their service areas. Of the three California utilities, SCE appears to have the largest number of customers who have voluntarily opted to participate in the load curtailment program. SCE's program consists mostly of large industrial, commercial, and agricultural customers who have agreed to temporarily curtail electricity usage during electrical emergencies in exchange for reduced rates. In times of severe power shortages, SCE calls upon the customers within the program—representing approximately 1,400 MW—to reduce their electrical usage. Typically, SCE activates its voluntary load curtailment program during times of a Stage Two emergency (indicating that power reserves have fallen below 5%). If power reserves fall below 2%, then the Cal-ISO could declare a Stage Three emergency, in which utilities could be directed to "drop load," necessitating involuntary rotating circuit outages for groups of customers across their service areas until sufficient reserve levels are achieved.

Moreover, as is often the case, California may be a case study for what appears to be renewed utility interest in the broad area of DSM programs. Again, it is important to note that DSM and voluntary load reduction programs are gaining momentum in regions where power reliability is a concern. Other areas that face few concerns about power supply or high electric prices may not find any renewed interest among utilities or customers in DSM programs. Secondly, as is the case with California, the jury is still out regarding whether or not interest in conservation will continue once the immediate threat of a power crisis has fully subsided.

Summary

The California energy crisis began to fizzle out in the summer of 2001 because the conditions that had brought the fundamental problems with the state's deregulated system subsided. The fact that these conditions subsided does not mean that the California energy crisis was solved during the summer of 2001, but rather the fact that the heat that caused the fundamental problems in the state to rise to a boiling point returned to a slow simmer. As will be discussed in the next chapter, California still faces a host of significant problems related to its restructured electricity market that have yet to be solved. If conditions once again take a turn for the worse, many observers fear that the California energy crisis of 2000-2001 could once again heat up and lead to problems for the state and its energy consumers. In other words, even though the California market has obviously stabilized, that does not mean that a year or two years from now the same severe crisis situation could not develop once again.

Chapter 10

Present Status: Where Will the California Energy Market Go From Here?

After almost a year of intense scrutiny that gained front-page headlines across the country, by the end of summer 2001 the California energy crisis had faded from the national news media. Other major events, especially the terrorist acts of September 11, took center stage and demanded the attention of the nation's federal leaders. In addition, the fact that the summer of 2001 came and went without a single power outage in California led many to believe that the energy crisis in the state had been resolved. That is a false impression. The California energy crisis, extreme as it was, clearly has subsided, but is far from over. Although the crisis may now be on slow simmer as compared to its previous boil, many of the fundamental issues that brought the crisis about in the first place have still yet to be solved.

As of December 2001, three months after the termination of direct access in California and almost one year after the state government stepped into its role as primary power purchaser, California legislators, regulators, and utilities still remain immersed in their attempts to resolve problematic issues. Further, California will soon head into another summer season, and if it proves to be significantly warmer than summer 2001, the state could very easily witness a repeat of some of the key problems of the California energy crisis. Issues that still need to be resolved include the following:

- The end of direct access
- The ongoing financial vulnerability of PG&E Co. and SCE
- The State of California's attempt to purchase the transmission infrastructure in the state
- Necessary transmission and generation upgrades
- Warnings of a possible electricity glut
- Possible refunds from out-of-state generators to the California utilities
- The risks associated with the state's efforts to re-regulate California's energy market

The End of Direct Access

As noted in the introduction to this book, on September 20, 2001, the CPUC jettisoned Californians' right to choose their power providers, tossing out the core of the state's disastrous bid to deregulate its electric industry. "Direct access is one-half of a failed and collapsed deregulation project," Commissioner Carl Wood said at the CPUC meeting that was more or less perceived as the funeral for direct access in the state. Commissioner Wood attributed the rest of the blame on the law's retail rate cap, which blocked IOUs from passing wholesale power costs down to their customers, incorrectly assuming wholesale prices would fall. In addition to gaining more control over the dysfunctional marketplace in the state, the CPUC's decision to end direct access was an admission that electric power customer choice had failed in California. The decision also essentially forbid energy customers in the state from seeking an alternate power provider other than their incumbent utility and, *de facto*, the state government.

Now that direct access has been terminated, the state has more certainty that its large customer base will remain intact and will ultimately share in the expected rate increases that are needed to pay off the state's outstanding debt for power purchases. One of the emerging "positives" out of this development is that, according to new reports, the termination of direct access in California should make it easier for the state to tap revenue from retail power sales needed to fund a record high $12.5 billion bond issue planned for later this year. The

bond will be used to repay the state Department of Water Resources for its
emergency power purchases. In fact, California State Treasurer Phil Angelides
disclosed that the CPUC had to terminate direct access in California before
the state would be able to sell the $12.5 billion in bonds to repay the state
treasury for the power it is buying on behalf of the three IOUs.

This is a very important point and needs to be clearly established. What
essentially drove the CPUC toward its decision to end direct access was a fear
that the state would be left holding a hefty bag of debt, with no recourse for
getting it recovered. Since the Department of Water Resources assumed the
role of operating as California's primary power purchasing agent, the state had
already racked up debts close to $11 billion by September 2001. It is the
state's intent to recoup this debt through rates affixed on energy customers
served by the three incumbent utilities. Yet, direct access would keep the door
open for large energy customers to sign deals with other power suppliers and
thus avoid having to pay any share of the state's energy tab, including purchases that had already been made. This could leave the state with a surplus
of costly power and few customers to pay for it. In addition, the state fears
that if small consumers are left to pay a disproportionately high amount of
the state's debt due to larger customers leaving the system, this would result
in a mass exodus from California, which would also have a negative impact
on the state's stability. This dynamic is known as the "fixed-cost death spiral,"
which the state is now trying to avoid. In the words of Loretta Lynch, the
CPUC's president, "We need to ensure a customer base...so that as the power
is purchased by the customer, the state is repaid."

Further, now that California direct access has been officially discontinued, the policy will most likely remain in place for at least 10 years, or
through the tenure of the long-term contracts that the state has signed with
various suppliers. As is clear from the history provided in this text, direct
access, which allows an end user to establish deals with private power
suppliers, had been in effect in California since the launch of deregulation in
the state in 1998. However, earlier in 2001, California passed a little-known
state law that put restrictions on customer choice in the state. Once the state
began purchasing electricity, it put restrictions in place such that customers
who had not yet switched from the incumbent utility were now prohibited

from switching. Thus, although direct access was officially terminated in the late summer of 2001, it had been dying a slow death for most of 2001.

Moreover, the effort to re-regulate California was not a new idea, but rather something that had been in the works for about a year. Nevertheless, despite the well-documented structural problems that are inherent in California's competitive model, the end of direct access in California was clearly rooted in the state's fear of competition and the opportunity for new suppliers to come in and take business away from the state. Further, the CPUC is reportedly considering, separately, whether it will later make its decision retroactive—something large consumers and alternative suppliers say would be unconstitutional.

Opposition to the CPUC's plan to end direct access had been intense since the Commission first began discussing this option. For instance, FERC Chairman Pat Wood had previously weighed in with his disagreement with the CPUC's expected vote to kill direct access. Wood argued that ending customer choice could lock Californians into higher energy prices (through the rate increases). Wood's position was that if direct access is not currently working in California, state officials should continue to modify their restructuring policies until direct access is successful (rather than abandoning it altogether). However, Wood backed off from interfering in the CPUC's vote, stating that the issue fell squarely under state jurisdiction.

Even the CPUC itself was divided on the issue. All of Gov. Gray Davis' appointees to the Commission—Commissioner Brown, Commissioner Carl Wood, and President Loretta Lynch—voted to end customer choice. The two remaining Republican appointees, Commissioner Bilas and Commissioner Duque, opposed it, arguing that more should be done to save a key feature of electric deregulation.

So, what does all of this mean? The CPUC has likely based its decision to end direct access on a belief that "desperate circumstances call for desperate measures." However, the desperate circumstances in which California still finds itself resulted from bad decisions made in the planning of the state's deregulated electric market, including decisions made by the CPUC. Moreover, the decision from the CPUC furthers the transfer of financial vulnerability from the incumbent utilities to the state of California, which creates a very precarious situation for the nation's largest energy market. In addition, electric power competition in the nation's largest state, which represents 20% of the country's economy, will now be blocked for at least the next decade. This potentially could have a very damaging effect on the level of new technologies and industries that would be brought into (or choose to remain in) the state of California. The total consequences of this measure remain to be seen, but it is clear that the CPUC's decision to end direct access is a huge milestone in the ongoing California energy debacle.

The CPUC vote may have signaled the official death of direct access in California, but the issue itself is far from dead. Almost immediately after the September 2001 vote, countermeasures contesting the CPUC's authority to terminate direct access in the state began to formulate. For instance, a coalition of California business groups and energy service providers have banded together to file a petition with the CPUC asking for a rehearing on its decision to suspend customer choice of electricity provider. The business and industry representatives argued that the Commission did not allow enough time for public input on the decision. "The commission has chosen to rush judgment on this issue, without hearings and without creating a proper evidentiary record," said Dan Douglass, an attorney for the Alliance for Retail Energy Markets (AReM). "AReM is prepared to take whatever legal action is necessary to make sure that direct access at least gets a fair chance at survival." The coalition said that it would consider a direct appeal the Supreme Court if the CPUC did not reconsider its decision on direct access.

The petition was signed by the following organizations: Association of California Water Agencies-Utility Service Agency, Western Power Trading Forum, AB&I Foundry, California Cast Metals Association, California League of Food Processors, California Retailers Association, Community College League of California, DDU Enterprises, Immanuel Industries, Lam Research, Spurr-Remac, Standard Metal Products, and Tricon Global Restaurants.

In accordance with the PUC Rules of Practice and Procedure, AReM and other business groups and concerned citizens filed an Application for Rehearing that identified the alleged errors of law made by the CPUC in its order suspending direct access. AReM cited the following primary reasons for why rehearing should be granted:

- The decision violates procedural due process guarantees
- The failure to hold hearings violates Public Utilities Code section 1708.5(f)
- The Commission's reliance on material outside the record violates due process
- The decision violates the Commerce Clause of the United States Constitution
- The threatened retroactivity is contrary to law and in excess of the Commission's authority
- The Commission acted contrary to law and in excess of its authority
- The decision's purported findings are not supported
- The Commission has impermissibly converted a ratemaking proceeding into a quasi-legislative proceeding

Consequently, the California Legislature has asked the CPUC to report back in January 2002 about what could be done to restore some sort of choice program without damaging the state budget or unfairly burdening any group of customers. Thus, debate is surely to continue over the direct access issue in California, and it remains unclear as of this writing whether or not the CPUC will be forced to rescind its decision and reinstate electric power competition in the state.

Financial Vulnerability of PG&E Co.

As discussed previously, PG&E Co. officially declared bankruptcy in April 2001. This by no means closed the book on the financial status of California's largest electric utility, but rather launched a very long and complicated process to bring the company back to some sort of financial solvency.

As discussed in the chapter covering the bankruptcy filing of PG&E Co., at the time of this writing the utility's parent is attempting to gain approval for a massive restructuring of its various units, with the obvious goal of protecting those subsidiaries that are still financially strong.

To recap, PG&E Corp.'s restructuring plan would essentially create two (possibly three) new companies: one for natural gas transmission, one for electrical transmission, and one that would own PG&E's hydroelectric and nuclear power plants. All of these companies would come under the corporate umbrella of PG&E Co.'s parent company, PG&E Corp., and would be operated separately from the electric and gas distribution utility. With 1,300 workers, the electric transmission company would operate PG&E's 18,500 miles of high-voltage power lines. The gas transmission company would employ 750 people and run 6,300 miles of natural gas pipelines in Northern and Central California. The generating arm of the company would operate hydroelectric dams and the Diablo Canyon Nuclear power plant, which together produce about 7,100 MW of electricity (about one-third of PG&E's total generation supplies). The generating arm would enter a 12-year contract with PG&E Co. to sell the energy from those plants for an average cost of five cents/kWh.

One of the current challenges for PG&E is that California state law limits the amount of money a utility can borrow. However, by moving the utility's transmission and generating assets into new subsidiaries, the new companies could borrow against the full value of those assets. It is PG&E's hope that the reorganization would help it to raise money to pay creditors $9.1 billion in cash and $4.1 billion in long-term loans. All creditors owed less than $100,000 would get cash payments for the full amount once the plan became effective, which PG&E said could be by the end of 2002. PG&E Co., the utility operation, would become a separate company with its own stock, under the reorganization plan. Keep in mind also that PG&E Corp. had previously established a ring fence approach to protect its National Energy Group division earlier in 2001.

The current reorganization plan could violate California state law that prohibits such asset transfers without state approval, which is one of the key concerns of the plan's critics. But PG&E Chairman Robert Glynn said his company believes the plan is legal because "a federal bankruptcy judge has the authority to overrule state law" if it is in the interest of creditors and the company. Mr. Glynn added that he doesn't believe the plan is subject to review by the CPUC or other state officials.

Nevertheless, state officials have weighed in with their disapproval of PG&E Corp.'s restructuring plan. For one, Gov. Gray Davis has stated that the plan gives the federal government (namely, FERC) too much control over the assets of PG&E Corp. (and, by the same token, reduces the amount of control that the state of California would have over these assets). Stating his position very clearly, Gov. Gray Davis said, "I am very wary of PG&E's proposal to transfer all of its generating capacity from a regulated environment to a non-regulated environment, which shifts oversight from the CPUC to the FERC." Further, Davis said that FERC has treated California ratepayers "shabbily" over the past 18 months and he believes that the CPUC should remain in control of PG&E Corp.'s assets to ensure a better deal for California ratepayers. In response to this criticism, PG&E Corp. says that its reorganization plan maintains the current regulatory authority for virtually all aspects of its business, and the CPUC will continue to regulate PG&E Co., including retail electric and natural gas rates.

Financial Vulnerability of SCE

The financial future of SCE is more difficult to predict than that of PG&E Co, if for nothing else because the plan that SCE will ultimately follow has yet to be fully revealed. It is true that SCE's credit has been reduced to junk status, but the company continues to make attempts to forge a rescue plan with the State of California that would keep it financially solvent, a plan that would clearly benefit both parties.

In early October 2001, SCE reached a settlement with the CPUC in a federal district court. The key elements of the plan included a rate freeze and an agreement by SCE to not pay shareholders a dividend until its existing debt is paid off. SCE's rates were raised by approximately 42% in early 2001 and will remain frozen through 2003 unless the utility pays off its debts sooner. In exchange, SCE agreed that it would use cash on hand and any revenue beyond what it needs to cover operating expenses to pay off its old debts; pay no dividends on its common stock through 2003 or until its back debts are fully paid; and drop a lawsuit against state regulators claiming the CPUC had violated federal law by failing to raise retail rates to reflect the underlying cost of wholesale power.

Under the agreement's terms, of the $3.3 billion debt Edison ran up when it couldn't meet its costs, about $1.2 billion would be covered by the company, the rest by continuing the present prices charged ratepayers until the end of 2003. The deal also essentially ended the maneuvering for a bailout of Edison by the California Legislature. Under the bailout terms, the company would have been able to raise rates to pay off bonds to cover the $3.3 billion debt.

Predictably, consumer groups blasted the settlement, calling it a bailout that froze SCE's electric prices at artificially high levels. Mike Florio, senior attorney for the consumer group TURN, argued that the settlement actually had "unhealthy implications for consumers." While the state legislature "refused to force consumers to pay for the bailout they were considering, the CPUC is insisting that small customers bear the brunt of Edison's problems," Florio said, noting that consumers would pay inflated rates indefinitely.

Gov. Gray Davis was supportive of the agreement, stating it "protected the public interest and would allow the state's second-largest utility to return to financial health," adding that he welcomed the CPUC's assurance this could be accomplished without raising electric rates.

However, the agreement between SCE and the CPUC hit a new snag in late October 2001 when a federal court blocked the agreement in order to provide consumer groups such as TURN adequate time to develop an appeal. The 9th U.S. Circuit Court of Appeals granted a consumer advocacy group TURN two weeks to argue against the settlement. However, on November 9, Judge Ronald Lew refused to further delay a settlement between SCE and state regulators designed to allow the utility to recover $3.3 billion of its debts and to keep it from bankruptcy. Judge Lew said delaying the deal, as requested by a consumer group, would risk harming the state's second-largest utility, its creditors and the public. Judge Lew, who approved the settlement on October 5, called the arguments for a stay advanced by TURN "repetitive" and "without merit." This was clearly a victory for SCE in the painstaking process of establishing a rescue plan for the utility with the state of California.

Nevertheless, the deal forged between the CPUC and SCE continues to meet with fierce opposition from other market participants. For instance generating company Mirant Corp. disclosed that it may still pursue an involuntary bankruptcy filing against SCE because it remains unsure the utility's recent settlement with state regulators will result in outstanding payment to various power generators, including Mirant. From SCE's perspective, the utility maintains that it ultimately will make good on all of the outstanding payments it owes to power generators. Edison CEO Stephen Frank responded to Mirant's threats with the following statement: "I don't know that there's any benefit to going the bankruptcy route when we clearly can pay them off in the near future."

Another interesting development has thrown a new wrench into the financial status of SCE. In August 2001, rumors began to emerge that City Light & Power Inc., a Denver-based company, was attempting a takeover of Edison International, including the SCE unit, the state's second-largest utility company. Later, City Light confirmed that it had been in negotiations to purchase Edison International for several weeks. However, representatives from financially-strapped Edison International continued to maintain that the company is not for sale and, even if it were, City Light would not be able to afford the acquisition. Nevertheless, what is particularly interesting about this rumored takeover is that City Light had previously attempted to buy the distribution assets of SCE through a partnership with Enron Corp., along with repeated efforts by the city of Long Beach to municipalize its electric system.

City Light's primary business model is in the operation of street lights for the city of Long Beach, which happens to be SCE's largest service area, with 175,000 customers reportedly generating $250 million a year for the utility in 1999. City Light has had a contract with Long Beach since the mid-1990s. However, the company also identifies itself as a "private utility development leader."

Despite the takeover rumors, City Light has no known partners or bank backing for the deal, and given its comparatively small size many analysts doubt it would be able to financially support the purchase (a point with which Edison International agrees). City Light is, however, being advised by Bear Stearns & Co. and Salomon Smith Barney.

Further, City Light claimed that its advisors were valuing Edison at $5.6 billion, or $17.18 per share, which represents a premium of about 21% over Edison's closing price on August 24, 2001. Edison has a market capitalization of about $4.6 billion, but of course has racked up over $3 billion in debt related to the high cost of wholesale power and rate freezes in effect in

California that preclude the company from passing on power costs to its customers. Investors expressed little interest in the prospect of a takeover of Edison; as of August 24, the day that the rumors of the takeover were confirmed by City Light, Edison's stock rose only two cents to close at $14.20. Given the financial complexities surrounding SCE, many analysts believe that the deal would be extremely difficult to navigate for a small company such as City Light.

In addition, Edison has consistently dismissed the reports out of hand, despite those City Light claims that negotiations have been in place for weeks. "It is fair to say that the idea apparently floated by a company called City Light & Power is without any merit," said Brian Bennett, Edison's vice president of external affairs, on August 24. "From what we know, City Light & Power is a small privately owned entity operating the street lights in Long Beach and has no visible financial resources. And, in any event, our company is not for sale."

As noted, this is not the first time that City Light & Power has made moves to purchase at least part of Edison International. In 1999, the company teamed with Enron Corp. to support a plan by the city manager of Long Beach to purchase Edison's distribution system. Thus, Edison's distribution assets are probably the primary attraction for City Light in this rumored deal. As noted, the company has acknowledged that it would attempt to purchase the totality of Edison International, but then possibly sell off some of the subsidiaries upon completion of the deal. Reading between the lines, City Light is arguably less interested in the generation portfolio managed by Edison Mission Energy than the distribution assets.

The city of Long Beach runs its own gas, water, and street lighting systems (through contracts with companies such as City Light) and purchases electricity from SCE under an unusual franchise agreement that in the past has given the city the option to purchase SCE's distribution assets. Most franchise agreements do not allow cities the option of purchasing a utility's distribution assets. Enron stepped in and formed a partnership with City Light, in which the two parties offered Long Beach a franchise fee of $8 million more a year than the city received from Edison. Under that scenario, had it been approved,

Long Beach would have bought out its contract with Edison and set up a municipal power distribution system run by Enron and City Light.

When SCE learned about the Enron and City Light deal, the utility began offering to triple the franchise fee amount because of concerns about losing the system that serves the city of Long Beach. The deal was ultimately scrapped when the city of Long Beach decided to negotiate a better deal directly with Enron, and City Light was essentially cut out of the negotiations. Ultimately, Long Beach accepted SCE's increased franchise fees and still receives power service from the utility. Nevertheless, the important point to note is that there is a history of City Light attempting to purchase at least the distribution assets of Southern California Edison and, with the new takeover attempt, has apparently resumed its interest in this proposition.

Consequently, these dynamics could very well set the stage for a takeover of Edison International by City Light or some other company. Presently, City Light is not considered a credible party to launch an expensive takeover attempt of Edison International, at least from investors. Yet, although drumming up the financial resources may be a challenge for the small company, the rumored takeover attempt could gain momentum if a state-endorsed rescue plan for Edison does not come to fruition. In addition, City Light has had some influential and financially strong partners in the past. Another company with designs to run California distribution assets and deeper pockets could very well step in as another player in this unfolding story.

Necessary Transmission and Generation Upgrades

As has been discussed in this book, in addition to power supply concerns, California has insufficient transmission capacity to meet the projected increase in generation. The strain that increased demand put on California's transmission system was one of the factors that contributed to the state's energy crisis. It is generally agreed that, at this point, considering the expected generation that is to come online in California in the next two to five years, transmission capacity in the state will not be sufficient to transport the necessary power. However, the three utilities in the state are taking a "wait-and-see" approach to determine if and where new power plants will be constructed in the state before devoting any capital to new transmission lines. As localized generation is also being increased, utilities believe that their costs for building new transmission lines must equal any benefits derived from the construction.

Nevertheless, according to the CEC, there are two transmission projects that have been formally proposed and are awaiting approval by the CPUC. Both projects probably won't begin until early 2002 and would not be completed until 2004 at the earliest. Any other projections on possible transmission upgrades in California would be speculative. The two transmission projects in California are: Path 15 and Valley Rainbow.

Path 15. In late May 2001, the Department of Energy directed the Western Area Power Authority to finish planning upgrades to a congested section of the Path 15 transmission lines in Central California. Congestion along Path 15, which transports electricity between the northern and southern parts of California, frequently contributes to power shortages in the state (particularly Northern California). PG&E Co. currently owns Path 15.

In October 2001, the federal government reached a deal with a coalition of energy companies to build a new transmission line under a $300-million project to relieve the chronic bottleneck along Path 15. The project, announced by Energy Secretary Spencer Abraham, is scheduled to begin in the spring of 2003 and could be completed by as early as summer 2004.

Amid a long list of uncertain ramifications of the Path 15 upgrade, there are some things that are known for sure. First, Path 15, which is owned by PG&E Co., is a 90-mile high-voltage section of transmission capacity that essentially carries power between Northern and Southern California. During the peak of the energy crisis in California, Path 15 became rather clogged and suffered from bottlenecks (points at which transmission could not transport high volumes of power). This pre-empted any transport of power from Southern California (and Arizona and Mexico) to the north, which resulted in rolling blackouts in San Francisco and other Northern California cities. As noted, these bottleneck problems remain and are particularly acute during times of peak power demand, which is obviously a concern as winter approaches.

The deficiency of the California transmission systems is arguably the last piece of the puzzle that remains toward bringing stability back to the state's energy market. Ironically, power supply is no longer seen as an imminent problem. New reports from the California ISO indicate that the state should have operating reserves of between 2,000 and 2,200 MW during the winter of 2001-2002, which have been deemed sufficient to stave off problems that plagued the previous winter season. However, the report on the state's transmission system is not as positive. If the Northwest continues to experience a drought, then Northern California will still need to import more power from the south. Heavy dependence on the strained Path 15 could once again cause reliability problems for California.

Thus, deficiencies along Path 15 can be seen as a symptom of larger problems that continue to plague California, and problems that have been the focus of both state and federal intervention for months, if not years. In fact, there is a history here that is worth noting, especially as it relates to federal / state jurisdiction. When President Bush released his national energy plan in April 2001, he directed the DOE to explore efforts that would relieve congestion along Path 15. Following through with this order, Energy Secretary Abraham issued a request for proposals that elicited responses from 13 companies interested in participating in the expansion of Path 15. The current contract, which will add a new line to the existing Path 15, is the result of that previous directive.

At present, it appears that PG&E Corp. and six other companies have signed on to financially support the upgrade. The six other companies that will spearhead the project are the Transmission Agency of Northern California, Trans-Elect, Inc., Kinder Morgan Power Co., PG&E National Energy Group, Williams Energy Marketing and Trading, and the Western Area Power Administration (WAPA), a federal agency within the DOE that sells electricity from water projects in 15 western states and operates 17,000 miles of transmission lines.

Presently, Path 15 connects the Bay Area in Northern California to two of the state's largest generation facilities—PG&E Co.'s Diablo Canyon nuclear facility and Duke Energy's Morro Bay natural gas power plant. The upgrade will reportedly include a new 500-kilovolt line that would boost transmission by about 1,500 MW from these two plants. Further, when the upgrade is finished in 2004, Path 15 should be able to carry a total of 5,400 MW of power.

Of course, the first question asked about the plan was: Who is going to pay for the upgrade? The way it appears at this writing is that the WAPA will retain 10% ownership of the new line for its role as project manager and for acquiring the rights from property owners to hold the power line. The remaining 90% will be owned by the private companies that are devoting different amounts of capital to fund the upgrade. The exact percentages of investment had yet to be defined as of this writing, but should be revealed in subsequent plans. Some have asked why these companies would want to participate in a transmission upgrade in California. The answer is that the companies will be able to recoup their investments through fees charged to use the new transmission line. This could result in a hefty payoff for the companies investing in the project, who presumably anticipate that California will remain a high-demand state through which huge amounts of power will need to be transported.

There are some potential conflicts that might arise with this upgrade plan. First, there is the issue of potential state/federal regulatory conflicts. Presently, state governments have jurisdiction over power-line siting. Thus, the CPUC had already been pursuing its own upgrade plan for Path 15 prior to Secretary Abraham's call for private-company bids. Specifically, the CPUC had directed PG&E Co., which it regulates, to upgrade the bottleneck spots along Path 15. As seems to be a growing occurrence, we may find that there is a state / federal jurisdictional dispute regarding which regulatory body has the authority to direct and manage transmission upgrades in California. Gov. Gray Davis and the CPUC certainly seem to believe that it is the state's prerogative, and have issued strong criticisms over the DOE's venture into this territory. However, in his energy plan, President Bush seemed to support a change in policy that would allow federal officials to obtain transmission rights of way as a way to increase national transmission capacity. If there is indeed a dispute between the CPUC and the DOE, it is not known if federal policy would supersede state policy in this area. In any event, representatives from Trans-Elect, one of the participating companies, have said that balancing the various interests among federal and state (and public and private) interests will be one of the key challenges for the upgrade project. Note that Trans-Elect was the company that had unsuccessfully attempted to purchase the transmission assets of SCE through much of 2001.

Also viewed by some as a potential conflict with the DOE's upgrade plan for Path 15 are the difficulties that could arise with private ownership of a major piece of California's transmission grid. The DOE-sponsored upgrade, unlike the one that the CPUC has directed, reportedly would not be subject to review or approval by either the California ISO or the CPUC. Further, because the companies participating in the upgrade will own a portion of the new line, these companies will be able to charge "tolls" to the power marketers that use the lines to transport power. This is most likely a point of contention for California officials, especially Gov. Gray Davis, who has sought state ownership of existing transmission lines and is not too keen on private-company ownership of generation facilities in the state. The positive side of this issue is that, according to the DOE, the bids from private contracts enable the upgrade of Path 15 to proceed without any increase to "either the taxpayers or ratepayers of California or the United States of America."

Valley-Rainbow Interconnect: In addition to Path 15, the CPUC is also proceeding with a new transmission line in Southern California, known as the Valley-Rainbow Interconnect transmission line. Although the timetable has been delayed until 2005, SDG&E is proceeding with a 500,000-volt transmission line that would connect SCE's Valley Substation near Romoland, CA, with a future substation to be located in the Rainbow, CA area, just south of the Riverside-San Diego county line. The new transmission line would provide a second interconnection between SDG&E's and SCE's transmission systems. The first interconnection is at the San Onofre Substation near San Clemente, CA. The project is designed to deliver approximately 1,000 MW. Studies have indicated that San Diego will exceed its reliable delivery capability in the summer of 2004 if the transmission line is not connected, so obviously SDG&E is anxious to receive approval from the CPUC and begin construction on the line. At this time, SDG&E expects to begin construction in 2002, with completion scheduled for the spring of 2004.

Generation Upgrades. As discussed in other chapters, in his series of executive orders, Gov. Gray Davis established a directive to get more power

plants running sooner and running harder in an attempt to avert any problems during the summer of 2001 and beyond. Gov. Gray Davis established
that the first 5,000 MW he expected to have online by July 2001 would
come from the following sources:

- Approximately 2,100 MW from new peaker units—which help the
  state through the busiest, or peak, hours of usage
- 1,630 MW from renewable energy sources, such as wind and solar
- 1,260 MW from large plants that are already under construction.

Those plants that meet these criteria reportedly would be able to apply
for permits from the CEC within 21 days, a process that normally has taken
a year in the state. Plants that do in fact make it online by July are eligible
for bonuses of as much as $1 million.

According to information from the CEC (current as of Nov. 1, 2001),
29 total power plant projects have been approved since the start of
deregulation in 1998, although not all of plants will be built. Three "major"
power plants, totaling 1,415 MW, have come on line in 2001 and are
producing electricity. Another 864 MW from "peaking" power plants were
scheduled to come on line by the end of September 2001. As of November
1, 2001, the CEC reported that 20 plants were under review, totaling 10,643
MW. These applications included simple and combined-cycle plants. Also as
of that date, the CEC had approved four emergency peaker units, under the
expedited 21-day approval schedule outlined by Gov. Gray Davis, which
totaled 435 MW.

Additional and current information about the locations of California
power plants approved by the CEC can be found at the Commission's Web
site (www.energy.ca.gov).

Warnings of an Electricity Glut

Ironically, concerns about California's supply shortage have now been
somewhat overshadowed by new reports of an energy glut. The main argument
goes like this: After a year of well-publicized concerns about a supply/demand

imbalance in many areas of the country, the energy industry presently finds itself in a boom cycle with regard to power production efforts.

In fact, the industry has quickly moved from one extreme to another, following years of virtually no power plant production in some areas with a current building binge. A report published in Barron's indicates that, as a whole, the industry plans to add as much as 290,000 MW of generating capacity over the next six years, which would represent an increase of about 40% over the current capacity of 760,000 MW. The Barron's report also suggests that the bulk of this new generation will likely come from merchant operators, which sell their output into unregulated markets. In some cases, this could increase the margin for capacity reserved for emergencies to an unnecessarily high 35%, up from a current 15% reserve. Coupled with a downward trend for forward prices for power, some analysts have warned that the accelerated increase in power supply could lead to a power glut, resulting in a huge amount of unused power.

It is important to note that this is mostly a national trend. For California specifically, the state has transitioned from having a deficient power supply to what should be adequate resources, at least in the near term. According to reports circulating in October 2001, new power plants and continued conservation should give the state enough of a margin to head off the soaring prices and rolling blackouts that plagued it a year ago, according to a forecast released by the California ISO. "In general, it looks like a pretty good forecast through the winter months into 2002," said Gregg Fishman, spokesman for the California ISO. In addition, it was reported that power-plant maintenance would be minimal during the 2001-2002 winter season compared with the previous year, and new power plants operating since September 30 would produce 2,231 MW in new supply, according to the California ISO report.

The California ISO expected demand for power to peak at 34,359 MW in October 2001 and about 32,000 MW from November 2001 through May 2002. That would leave operating reserves of between 2,000 and 2,200 MW—roughly a 7% reserve margin—the minimum amount required by the Western Systems Coordinating Council, which maintains grid reliability throughout the West. The California ISO said the reserve margin would be enough to avert the disaster the state experienced during the winter of 2000-2001.

In November 2001, the CEC issued a forecast that the State would have adequate supplies of electricity to meet the demand of summer 2002, as long as planned power plants are built and current levels of conservation continue. The CEC did acknowledge that July 2002 could be a tight month for the state, as a supply surplus of only about 340 MW (including new generation of 4,000 MW) is expected to be online by that time. In the case of an emergency declared by the California ISO, approximately 1,700 MW of electricity should be available from large users participating in interruptible/emergency demand responsive programs. Much of the positive projections for the summer of 2002 rest on continuing conservation efforts on the part of California energy customers, something that is not easily predicted. If Californians perceive that the energy crisis is over and there is little need to conserve energy in the same manner as they did during the course of 2001, this could have an impact on available resources in 2002 and beyond.

Refund Issues

The issue of whether or not power generators that sold power to the California utilities should make refunds to the state is one of the main unresolved issues from the California energy crisis. The issue has its origins in the previous determinations by FERC that market rules and structure for wholesale sales of electricity in California were "flawed" and had caused "unjust and unreasonable" rates.

Based on the assumption that wholesale power prices in California had been unjust and unreasonable, in July 2001 FERC appointed an administrative law judge to investigate the matter and address whether or not power generators that had previously sold power into the California market owed refunds back to the state. State officials, including Gov. Gray Davis, repeatedly demanded that generators owed the state approximately $8.9 billion in wholesale electricity refunds, based on data compiled in a reported conducted by the California ISO. After two weeks of negotiations, Curtis Wagner, chief administrative law judge with FERC, announced that refunds would be returned to

California, and that the vast majority of the methodology presented by the California delegation would be adopted in his recommendation.

In the clash of titans that the war between power suppliers and the state of California had become through most of 2001, it was no surprise that a settlement could not be reached, despite two weeks of intense negotiations and concessions made on both sides. Although the state of California had agreed in principle to various non-cash forms of payment from the generators, negotiations stalled due to a huge philosophical gulf that existed between the two parties and a seemingly irreconcilable quantitative dispute over the amount of money that was owed. Gov. Gray Davis would not budge regarding his claim that generators owed the state nearly $9 billion. Most of the generators consented to paying some amount of refund, but claimed that the figure should be no more than $1 billion. The wide range between the two amounts was what caused the negotiations to hit a brick wall and caused Judge Wagner to assume responsibility for a final recommendation to FERC. The stakes in this decision were huge, as Judge Wagner's refund plan (and FERC's subsequent ruling based on its own conclusions) would undoubtedly set a precedent for how wholesale power was bought and sold in U.S. markets that were subsequently deregulated.

Although much of this case remained in contention, some key factors were fairly undisputed. For instance, FERC had established that it was only seeking to establish a ruling based on three key areas: refunds of overcharges, long-term power contracts, and debts owed to generators by California utilities. Also, it was clear that the time period for which power generators may ultimately have to issue refunds was October 2000 through May 2001. Any overcharging that might have occurred outside of this timeframe was beyond the scope of this investigation and would not be included in any refunds made by the generators. The other thing that was known was the list of generating companies that were singled out in the FERC investigations. The cast of players included the most familiar companies: AES (NYSE: AES), Calpine (CPN), Duke Energy (DUK), Dynegy (DYN), Enron (ENE), Reliant (REI), Mirant (MIR), and Williams (WMB), which all were subject to potential refunds.

Regarding the dispute, Gov. Davis maintained that the $8.9 billion in refunds, a figure that was calculated by the California ISO, was an accurate amount and one that was non-negotiable. Gov. Davis acknowledged that the figure included potential overcharges from government-run operations such as the LADWP and BC Hydro, which FERC does not regulate. If the municipal utilities were not included in the equation, California officials alleged that the power generators listed above collectively owed the state around $6 billion. Yet, when considering other Western states besides California that may also seek refunds from the power generators, the potential refund claimed by the states could reach as high as $15 billion. Records provided to Judge Wagner by California officials indicated that in January 2001, the first month in which the Department of Water Resources started to serve as the state's power purchaser, California spent about $332/MWh for power on the wholesale spot market. In the latest news coming from Judge Wagner, it appears that he is leaning toward an acceptance of the data presented by the California officials, which could mean a big win for the state and a potentially big loss for the generators.

In response, generators maintained that the $8.9 billion figure had not been substantiated and is far beyond the reported $1 billion that they would be willing to refund to California as a collective group. In fact, Judge Wagner previously had also gone on record as stating that the figure thrown out by California officials was too high. In addition, attorneys for the generators claimed that it is "ludicrous" for the state to demand refunds of this magnitude when generators were still owed billions of dollars from California's financially strapped utilities (from the days before the state took over as power purchaser). The reticence on the part of the generators stemmed in part from a belief that they are not totally responsible for the high cost of wholesale power that was common during the timeframe in question. Rather, generators claimed, high prices resulted from a combination of high natural gas prices, restrictions in the California market that blocked adequate power supply in the state, and other market conditions. However, it is important to remember that companies such as Duke, Dynegy, Enron, Reliant, and Williams are among the leading gas traders in the United States and thus often control the price of natural gas.

Judge Wagner disclosed on July 9, 2001, that power generators had offered a settlement of $716 million, which California officials flatly refused.

Of that amount, reportedly $510 million was offered by the five biggest generators—Duke, Dynegy, Mirant, Reliant, and Williams. California's rejection of the offer signaled the end of the negotiations.

The actual dollar amount that the state of California may have been overcharged remains ambiguous because, up to this point, power generators have refused to provide data showing their actual cost of producing power, as this represents competitive information. Rather, all of the parties involved in the negotiations have made presentations to Judge Wagner, and naturally have made the best case to support their various positions. This could all change if the dispute makes its way to a full-blown federal trial, as power generators (and state officials, for that matter) might be forced to turn over proprietary operational data.

Another obstacle that has prevented any sort of resolution in this case is the fact that generators want, in exchange for the refunds, a commitment from the state that it will call off all pending civil cases and related investigations against the generators. In fact, some of the generators in question have reportedly said that they will make some level of a refund, but only on the condition that state officials sign documents agreeing to abandon any civil lawsuits against the generators and their executives.

Ironically, after all the public sparring that has taken place between Gov. Davis and the power generators, at times the parties seemed to move closer toward a resolution than at any other time since California utilities spiraled into debt a year ago. In a significant step toward compromise, Gov. Davis said that although the $8.9 billion demanded by the state is non-negotiable, various options besides a cash payment might be acceptable. These other options include adjustments to long-term power contracts that the governor signed earlier this year, contracting for additional power at below-market rates, or credits to energy companies for unpaid bills. In other words, Davis would be willing to work with generators that have already signed contracts with the state to renegotiate a better deal that would be applied to the possible refunds.

In October 2001, FERC did issue a directive on refunds, but the issue has yet to fully resolved. Specifically, FERC ordered Dynegy Corp., Mirant Corp., Williams Cos., and Reliant Energy to give the refunds to the state of California. Though FERC did not specify how much money each company should be prepared to give back, the operators of the California electricity grid estimate that in all, they were overcharged $260,000 by power sellers in July 2001. The refunds were based on FERC's previously established wholesale price ceiling and the fact that the companies had not provided sufficient justification for charging prices above that price ceiling. The FERC order concluded that the justifications submitted by the four companies for electricity sold in July were either not filed on time or were unsubstantiated. FERC found that Reliant Energy, for example, did not offer sufficient detail about the price of the natural gas that the company purchases to run its Southern California power plants.

Of the four companies, only Williams released the amount of money it is being asked to refund. Paula Hall-Collins, spokeswoman for the Tulsa-based company, said that $30,000 worth of electricity sales were subject to the FERC rebate order for both June and July. Williams said it would ask for a rehearing on the issue because the company opposed the method federal regulators used to calculate the price ceiling. Dynegy spokesman Steve Stengel said his company also would ask FERC to reconsider its order.

Keep in mind also that the Department of Water Resources also owes money to the power suppliers. In November 2001, FERC ordered the California ISO to produce a plan within three months that would outline how the Department of Water Resources would pay some $1.2 billion in past-due energy bills. FERC directed the California ISO to send an invoice within 15 days to the Department of Water Resources for all transactions it made on behalf of the state utilities dating back to the beginning of 2001. This order will no doubt spark objections from the Department of Water Resources, which has essentially refused to make the outstanding payments until it receives detailed information on bids and prices for previous transactions from the California ISO. The California ISO has repeatedly claimed that it cannot turn over such price information, which power generators and suppliers maintain is proprietary.

Consequently, the refund issue is far from over and will undoubtedly carry into 2002 and beyond. Keep in mind that the refunds FERC ordered the four generators to pay in October 2001 were only for the June and July 2001 time periods. Other refunds for other time periods could subsequently be ordered by FERC. There undoubtedly will continue to be ongoing lawsuits and appeals launched by the generators ordered to give refunds, as many of these companies still have not been paid for previous power sales and will refuse to make any refund payments until previous transactions are resolved.

Risks Associated with State's Role in California's Energy Market/Bond Issues

When Gov. Gray Davis directed the California state government to step into the role of power purchaser in January 2001, the State of California clearly began to assume a whole new set of risks to which it had not previously been exposed. As noted, typically, the power contracts between the Department of Water Resources and power generators, and signed by Gov. Gray Davis, are in place for 10 years and were based on higher-than-cost-to-service rates. Some reports have indicated that the long-term contracts make the state liable for $43 billion in power payments over the next decade. At the time the contracts were signed, Gov. Davis felt pressure to ensure that California had a reliable supply of power, but he reportedly has been criticized for locking the state into expensive, long-term agreements and is eager to renegotiate them. While Gov. Davis is flexible on the various forms that the refunds could take, he has said, "they have to net out close to $8.9 billion." However, some of the power generators that established contracts with the state say re-negotiation was not possible as they had already locked themselves into deals with natural gas suppliers.

Unfortunately, these long-term contracts did not take into account the decline in demand from a recession that might occur both in the state and nationally, which would drive down wholesale prices. In other words, the contracts signed by the state of California were based on wholesale market prices as of early- to mid-2001, a time of comparatively prices. As prices undoubtedly will fall in the subsequent years, due to increased supply and ongoing conservation efforts, the state of California will be locked into comparatively high prices for power. Realizing its mistake, the California state government has launched ongoing efforts to renegotiate the various long-term contracts it signed with power generators.

Many electricity analysts warned all along that it was folly to enter into long-term contracts for a short-term crisis. However, now that the contracts have been signed, there may be little that the state of California can do to revise the terms of the contracts. As of late October 2001, California Senate President Pro Tem John Burton called for Gov. Gray Davis to renegotiate California's long-term energy contracts for fear the state otherwise will be stuck "paying too high a price for too much power for too long a time."

Gov. Davis' spokesman Steve Maviglio said the state appears to be bound by the contracts negotiated at the height of the state's energy crisis, adding that criticism now may be shortsighted. Specifically, the state appears to be locked in to spending more than $40 billion over the next 20 years buying power under contracts regarded by some lawmakers, the CPUC, and other groups as grossly overpriced and possibly tainted with conflicts of interest. Maviglio contended the contracts helped end the crisis by driving down electricity prices, and noted the contracts were well below spot prices in effect at the time. Also, many are tied to construction of new power plants that still will be needed as the state's economy and population grows, Maviglio said.

However, Sen. Burton, D-San Francisco, said the state was forced to make "unreasonable and inflexible concessions," locking in higher prices long-term in exchange for a lower short-term cost. One of those technical contract provisions now endangers the state's ability to issue long-term bonds to repay the state's treasury the $6.1 billion the state spent to buy power in the summer of 2001 on behalf of three cash-strapped utilities, Burton said. That, in turn, may hurt a state budget already affected by a declining economy. Burton argued the state should push to renegotiate because the contracts "were negotiated under economic duress." Loretta Lynch, president of the CPUC, said that state would have to renegotiate about 50 contracts to spare Californians from paying high electricity prices for years to come.

The bond issue is also significant. The other issue that remains unresolved at this point is the issuance of $12 billion in state-backed bonds, which had been previously pushed by Gov. Gray Davis but was not included in the CPUC/SCE settlement. The bonds would have been used to finance electricity purchases made by the Department of Water Resources, but the issuance of bonds was defeated in a 4-1 vote by the CPUC. The commission apparently was concerned about any efforts to diminish its own authority to regulate power prices. However, at this point the issue is still unresolved, and it keeps the state of California in a precarious position in terms of how it will continue to finance expensive power purchases. The state of California must now go back to the drawing board to find a way to sell the record $12.5-billion bond, although the CPUC's refusal to take action on this issue means there is presently no schedule for the bond issuance and, consequently, great uncertainty about the State of California's own financial stability regarding its involvement in the electric power business.

At the time of this writing, the California Legislature was working to expedite a $12.5-billion bond sale to finance power purchases that the state continues to make to provide power for the cash-strapped utilities. The state plans to sell the bonds, the largest municipal debt offering in U.S. history, to repay the state's general fund for a $6.1 billion loan and pay for other power costs. The state Department of Water Resources has reportedly spent $10.4 billion buying power (as of October 2001) on behalf of utilities since January 2001. However, a major point of contention surrounding the bonds is the fact that power generators, who are in long-term contracts with the state, believe that their contracts entitle them to first claim on the bond proceeds. Given the fact that many of the power generators have yet to be fully compensated for previous power sales made to California IOUs and the Department of Water Resources, this could become another dilemma for the state if the bonds are ultimately approved.

New Plan to Renegotiate Long-Term Contracts and Build New Power Plants

After months of defending the $43 billion worth of long-term electricity contracts he helped negotiate on behalf of the state, S. David Freeman suggested for the first time in 2001 that the contracts be renegotiated, perhaps through the California Power and Conservation Financing Authority, a new public power agency he now chairs. "There seems to be pretty general agreement that these contracts need to be renegotiated," said Freeman, noting that critics of the contracts include Gov. Gray Davis, the president of the California Public Utilities Commission, and the leader of the State Senate. Freeman said he is still proud of his work negotiating the contracts with companies Gov. Davis labeled at the time as gougers and pirates, but California's energy picture has vastly changed since January2001.

At issue presently are some 54 long-term contracts that the state of California, through its Department of Water Resources, signed with power generators such as Calpine (NYSE: CPN), Duke (NYSE: DUK), Mirant (NYSE: MIR), and Williams (NYSE: WMB), to name just a few, back in early 2001, at a time at which wholesale power prices were still running at very high levels. One of the primary benefits of the contracts was that it reduced the state's reliance on the volatile spot market, where prices had soared as high as $500/MWh. As a whole, the contracts are worth about $43 billion and have a lifespan of 10 years or more. Some might argue that Gov. Gray Davis, who led the effort for the state to assume the role of power purchaser, felt pressure to sign the contracts at that time, due to the uncertainty surrounding the financial solvency of PG&E Co. and SCE in particular. However, critics argue that the contracts locked the state into wholesale power costs when prices were the highest.

Many of the details of the contracts signed by the state are proprietary, but there are some interesting details that can be gleaned. First, a good number of the contracts lock the state into buying power at various times, including those of low demand (such as the morning). This leaves the state with a surplus of power that it does not need, which it in turn has been forced to sell at a loss. We also know that, as a general observation, the state bought power under the long-term contracts at an average price of $75/MWh. That same power reportedly will sell for only $16/MWh in 2002.

As noted, Freeman, who previously managed the municipal utility known as the Los Angeles Department of Water and Power, recently assumed the management post of the new California Power and Conservation Financing Authority at the request of Gov. Gray Davis. The agency was charged with quelling the extreme situation that California faced over the last year, including soaring power prices and blackouts. Looking beyond the immediate problems that have subsided, Freeman's new plan calls for a way to renegotiate the existing contracts and increase generation supply in the state at the same time.

The Freeman plan can be summarized rather easily. As a state agency, the Power Authority could sell up to $4 billion in revenue bonds, which would be guaranteed by energy sales, to lease, build, or buy power plants. Consequently, the state, which can borrow money at below-market rates, is in a position to build new plants more cheaply than private companies could. As a carrot to entice the renegotiation of the long-term contracts, the state could offer generating companies a financial incentive to build new power plants in the state. In other words, the state would carry the investment for the costs of the new plants, alleviating pressure on the private companies to provide a 20% return to their shareholders. Note that most of the generating companies involved in long-term contracts with California are committed to building new power plants anyway. Some reports I've seen indicate that 70% of the 54 contracts that the state has signed include clauses that require the generating companies to build new power plants in the state. However, under normal circumstances, the expense of building the new plants would be financed by the generating companies and could cost hundreds of millions of dollars. Thus, in return for the financial incentive, the same generating companies would agree to renegotiate the terms of their long-term contracts with the state of California, presumably based on current market conditions. Note that the renegotiation could include cutting the prices in the contracts, or providing the state with more flexibility on the timing and quantity of electricity that must be purchased.

The word from California is that most state officials think this plan has some legs. Nevertheless, word of the plan comes on the heels of claims that California is presently suffering from an energy glut (unused electricity) that may end up costing ratepayers as much as $3.9 billion over the next decade. The reason for the surplus power is that Californians have increased conservation efforts, which brought demand down, a condition that was maintained by comparatively moderate weather trends. The end result is that the state apparently bought far more power than it needed to meet the needs of the customers served by the three IOUs, and the state unfortunately cannot sell the excess power elsewhere and gain a profit. For instance, according to a report by the Department of Water Resources, in one three-month, low-usage period expected in the spring of 2002, 57% of the power for which the state has contracted will have to be sold at a loss of close to 80 cents on the dollar, ultimately costing utility customers as much as $193 million. The same report indicates that the power surplus in the state will reach its peak in 2004 and then gradually decline through 2010.

Consequently, despite the financial advantages of Freeman's plan, the logistics of getting new power plants approved in the state in light of the apparent energy glut may be an impediment to the renegotiation strategy. In addition, some of the companies involved in the contracts with the state already own a large amount of generation capacity in the state and may not be easily convinced to build new plants right away. For instance, Calpine Corp., which is one of the companies that has signed a long-term deal with the state of California, responded to the plan as saying that it "wouldn't be anything that Calpine would use." The company reportedly has finished three new power plants since June 2001 and would not be enticed by the financing incentive that the state is orchestrating. Further, the state's desire to renegotiate terms of the contracts is not a new concept. Ever since wholesale prices began to drop earlier this year, the renegotiation debate has been a fixture of state legislative and regulatory proceedings. However, since the onset of the talks, a good number of the generating companies have maintained that re-negotiation would not be possible as they had already locked themselves into deals with natural gas suppliers.

Thus, one concern is that the apparent energy glut in California will discourage further development of new generation and renewable energy sources, at least in the near term, which could set the state up for another dangerous boom-and-bust cycle down the road. In addition, those companies that already have long-term contracts with the state (such as Calpine) may not be enticed by the financing incentive, and those companies that don't have long-term contracts with the state will have no incentive at all to build new plants in the state. All of this could create a situation in which California has too much power over the next few years and then will find itself in another shortage situation 10 or more years down the line.

Freeman is apparently also pushing the state to once again make an attempt to take over some of PG&E Co.'s physical assets. Instead of its transmission lines, however, the state now is examining its opportunity to buy the hydroelectric generation network (including dams and powerhouses) owned by the state's largest electric utility. Note that under its restructuring plan that has been submitted for regulatory approval, PG&E Co. would split from its parent, PG&E Corp., and transfer generating and electric and gas transmission assets to form three new companies, which would fall under the jurisdiction of the FERC. State regulators do not like this plan, and thus are seeking a way to retain control over the generating and transmission assets of PG&E Co. Putting the hydroelectric assets under the State Power Authority would keep the assets under state control.

Moreover, even though California is rethinking certain choices it made almost a year ago that directly entrenched the state government in the energy market, state officials still seem to want to gain control over fundamental parts of the state's energy infrastructure. Remember that Gov. Gray Davis spent much of the last year attempting unsuccessfully to negotiate deals with SCE and SDG&E for the purchase of the utilities' transmission networks. PG&E Co. never was interested in selling its transmission assets, which it believed the state's offer grossly undervalued. One of the main reasons that these attempts by the governor were unsuccessful was concern by state legislators, who argued the state was ill equipped to assume operation of the complex transmission networks. Nevertheless, in addition to renegotiating the long-term contracts, the state is still pursuing at least the hydroelectric generation assets of PG&E.

Conclusion

Perhaps the most important message to conclude this discussion of the California energy crisis is that the state's problems related to the restructuring of its electricity market are far from over. In fact, one of the distinct challenges in constructing a thorough analysis of electric restructuring in California is finding a point at which to stop writing about this fluid story. Even at the end of 2001, nearly four years after direct access started in California and four months after direct access was pronounced dead, news continues to emerge about the state's efforts to repair its struggling energy market. Many critics believe that what has taken place up to this point is a series of "band-aids" that have been placed on the market in an attempt to minimize the bleeding from the failed deregulation experiment. Moving forward, most agree that California needs to find some long-term solutions to cure its problems. However, what those long-term solutions are is a question that remains cast in shadows and the subject of intense debate.

Clearly, what California has learned now is that its original competitive market design did not work. That is at least one point on which all participants have found agreement. Yet, the reality is that by initiating the steps to dismantle its regulated monopoly energy system, the state of California presently finds itself in a difficult limbo period. Put bluntly, the state can't go back to the way things used to be, but it still has not figured out how to move forward either. Thus, California finds that its energy market is still in a state of flux as it continues to gauge the impact from previous decisions, with the state's government currently bearing the greatest share of the responsibility related to this uncertainty.

Meanwhile, larger market forces continue to cast doubt over the status of electric power deregulation across the United States, a path that California certainly blazed with less-than-successful results. At the time of this writing, Houston-based Enron Corp., one of the energy suppliers often singled out by Gov. Davis in his crusade against high wholesale prices in California, was closing out the year 2001 in the midst of great financial instability and a bankruptcy filing of its own. Enron's financial problems had little to do with the California market, but instead were the result of losses associated with the company's investments in non-core businesses such as telecommunications and water. Nevertheless, Enron certainly had been one of the most vocal proponents of electric deregulation and had in fact been quite instrumental in encouraging electric power competition in California. The fact that Enron became financially weakened through its expansion efforts made possible by a competitive energy market led many to believe at the end of 2001 that faith in electric power deregulation on a national level would be severely damaged.

There is certainly evidence that a good number of states have observed the California experience with deregulation and made the decision that proceeding with electric power competition may not be a prudent move at this juncture. For instance, states such as Arkansas, Oklahoma, Iowa, and New Mexico (to name just a few) have all recently cited the California experience as a main factor in their decisions to delay or terminate altogether any plans to open their own markets to electric power competition. Granted, these states tend to have lower-than-average electricity rates, so competition has not been as high a priority for large industrial customers in these states as it was in California. However, it is a fact that at the close of 2001, the United States is equally divided between states that have approved some sort of restructuring plan and those that have not. Among the states that have not developed a restructuring plan—and even among those states that have—California is still viewed as an extreme case of the risks that a state can encounter when proceeding with electric competition.

In contrast, another philosophical argument suggests that electric deregulation is inevitable and should not be impacted by the California experience (or Enron's problems for that matter). From a national perspective, the FERC appears to be committed to opening electricity markets to competition, despite the problems in California and the bankruptcy of Enron. FERC continues to move forward with formulating policy for the construction of regional transmission organizations (RTO), the entities that will oversee wholesale activity across transmission grids and theoretically enable competition among power suppliers. In addition, as noted in this text, Texas and Pennsylvania represent examples of states for which great hope about electric power competition is still reserved. One positive is that many states have looked upon the California experience and recognized it for it is— an anomaly. In other words, California developed a very unique model for electric power competition that contained fundamental flaws. It does not represent a template for how deregulated markets might function in other states or how other states might choose to structure their own competitive electric markets.

However, this text has been about California and not about deregulation efforts across the United States as a whole. The unfortunate fact of the matter is that California made some very serious mistakes in formulating its model for electric power competition, and those mistakes continue to put the state into a very vulnerable position even after direct access has been officially discontinued. Making matters potentially even worse, state officials have responded to the failed experiment of deregulation by deciding to assume even greater involvement and control in the energy infrastructure of the state than they had under the monopoly system.

Specifically, instead of developing a repair policy that would allow the IOUs to once again own and control their own generation supply, the state government in California has assumed the responsibility for buying power for the utilities and possibly gaining ownership of their transmission and any remaining generation assets they may have left. It is perhaps too early to determine any financial repercussions that this choice on the part of state officials will have for California and its residents. However, it cannot be denied that this choice has its own set of unique risks that may be just as disastrous as the original model for direct access.

Unfortunately, the state of California—as the first state in the nation to enact electric power competition—ventured into previously uncharted territory and encountered mammoth problems as a result of this exploration. As a trailblazer, California paved the way for other states to make more educated decisions about how to offer electric power choice in their own energy markets. For this alone, the California deregulation experiment was not conducted in vain. However, from its own perspective, it may take a good number of years for California to find a way to repair the array of mistakes that were made in implementing electric power choice and minimize the ongoing impact of its attempts to re-regulate its energy market.

Index

A

B

Bank of America, 64

Bank of New York, 64

Bankruptcy (Enron), 55

Bankruptcy (PG&E Co.), xix, 50-51, 62-64, 162: announcement, 62-63; non-utility protection, 63-64

BC Hydro, 58, 177

Biomass, 3

Blackouts (power), 25-28, 30, 44-46, 145-147

Blame game (PG&E Co.), 64-65

Blue Book, 6-8, 10

Bonds, 73, 76-78, 180-182: financial risk, 180-182

C

Cabrillo I, 91

California Assembly, xxiv-xxvii, 109, 115-116: California State Legislature website, xxiv

California Codes/laws, xxvii

California Constitution, xxvii

California deregulation, 5-16: CPUC policy, 6-10; transition process, 8; California ISO, 8-9; California PX, 9-10; Assembly Bill 1890, 10-12; federal measures, 13; new frontier, 14-16

California Energy Commission (CEC), xxii, 3, 18, 93, 95, 146173, 175: website, 95, 173

California energy crisis, 17-37, 39-60, 145-155: causes/reasons, 17-37; impacts, 39-60; dissipation, 145-155

California energy market, 89, 157-186

California Environmental Quality Act, xxiii

California Independent System Operator (ISO), xx-xxi, 8-10, 14-17, 57-58, 96-99, 102-103, 110, 129, 132-133, 135, 139, 145-148, 154, 170, 172, 174-175, 177, 179

California Power and Conservation Financing Authority, 58-59, 182-184

California Power Exchange (PX), xxi, 9-10, 14-15, 17-20, 27-28, 33, 40, 46-47, 61, 109, 128, 139

California Public Utilities Commission (CPUC), xi-xv, xxii-xxiv, 2-3, 6-12, 14-17, 20, 27, 43, 67-71, 76-77, 82-89, 96-100, 113, 117, 125, 158-162, 164-165, 171-172, 181-183: policy, 6-12

California State Legislature website, xxiv

Calpine Corp., xxx, 52-53, 91, 93, 113, 176, 183, 185

Cambridge Energy Research Associates, 94

Causes/reasons (energy crisis), 17-37: wholesale electricity market flawed, 19-21; power supply/demand, 19, 21-23; power imports decreased, 19, 23-25; in-state generation off-line, 19, 25-28; wholesale prices rising, 19, 28-29; transmission grid strained, 19, 30; weather patterns, 19, 31-32; market manipulation by power generators, 19, 32-37

Chaptered bills, xxvii

City Light & Power Inc., 166-168

Coal, 48

Cogeneration, 3

Colorado, 132

Committee hearings, xxv-xxvi

Competition transition charge (CTC), 11-12, 41

G

F

H

J

K

I

L

Other titles offered by PennWell...

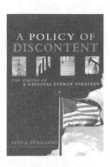

A Policy of Discontent:
The Making of a National
Energy Strategy
by Vito A. Stagliano
446 pages, hardcover
$39.95 US/CAN
$54.95 Intl
ISBN: 0-87814-817-5

Managing Energy Risk: A
Nontechnical Guide to
Markets and Trading
by John Wengler
393 pages, hardcover
$64.95 US/CAN
$79.95 Intl
ISBN: 0-87814-794-2

Fundamentals of Trading
Energy Futures &
Options, 2nd Edition
by Steven Errera and
Stewart Brown
247 pages, hardcover
$64.95 US/CAN
$79.95 Intl
ISBN: 0-87814-836-1

Creating Competitive
Power Markets: The PJM
Model
by Jeremiah D. Lambert
236 pages, hardcover
$64.95 US/CAN
$79.95 Intl
ISBN: 0-87814-791-8

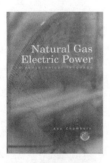

Natural Gas & Electric
Power in Nontechnical
Language
by Ann Chambers
258 pages, hardcover
$64.95 US/CAN
$79.95 Intl
ISBN: 0-87814-761-6

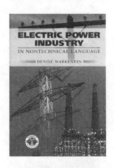

Electric Power Industry in
Nontechnical Language
by Denise Warkentin
239 pages, hardcover
$64.95 US/CAN
$79.95 Intl
ISBN: 0-87814-719-5

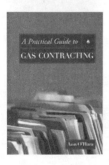

A Practical Guide to Gas
Contracting
by Ann O'Hara
467 pages, hardcover
$64.95 US/CAN
$79.95 Intl
ISBN: 0-87814-764-0

The New Rules: A Guide
to Electric Market
Regulation
by Steven Ferrey
370 pages, hardcover
$64.95 US/CAN
$79.95 Intl
ISBN: 0-87814-790-X

TO PURCHASE A PENNWELL BOOK...

- Visit our online store www.pennwell-store.com, or
- Call 1.800.752.9764 (US) or +1.918.831.9421 (Intl), or
- Fax 1.877.218.1348 (US) or +1.918.831.9555 (Intl)